Durable Pleasures

By the same author

THE *TRUE* SOUND OF MUSIC
A practical guide to sound equipment for the home

THE WALTZ KINGS
Johann Strauss, Father and Son, and Their Romantic Age

WILLIAM PENN
Apostle of Dissent

Durable Pleasures

A PRACTICAL GUIDE TO BETTER TAPE RECORDING

by Hans Fantel

A Sunrise Book

E. P. Dutton & Company, Inc. | New York | 1976

First Edition
10 9 8 7 6 5 4 3 2 1

Dutton-Sunrise, Inc., a subsidiary of E. P. Dutton & Co., Inc.
Published simultaneously in Canada by Clarke, Irwin & Company Limited, Toronto and Vancouver

Library of Congress Cataloging in Publication Data

Fantel, Hans.
Durable pleasures.

"A Sunrise book."
Includes index.
1. Magnetic recorders and recording—Amateurs' manuals. I. Title.
TK9967.F36 1976 621.389'32 75–17730
ISBN 0–87690–185–2

for Shea

Acknowledgments

The section of this book entitled "The Nature of Sound" has appeared in somewhat different form in *Stereo Review*, where I served as associate editor for several years. I am grateful to Mr. William Anderson, the editor of *Stereo Review*, for permission to adapt the material for use in this book.

H.F.

CONTENTS

LIST OF ILLUSTRATIONS

HOW TO USE THIS BOOK

You don't have to read this book from beginning to end. You can, if you want to. In that case, you will find first a survey of the various types of tape recorders available, followed by an explanation of how they work and what determines their quality, and finally a series of practical suggestions for making different kinds of recordings.

But perhaps you are interested in only one of these topics. If so, skip around and pick out just the chapters with the information you need. To help you find your way, here is the basic plan:

Chapters 1 through 6 of this book deal with choosing your equipment. Chapters 7 through 11 deal with using it. In that section you can get detailed, practical information on how to record off the air with top fidelity, copying phonograph records, using your recorder in business or school, taping on the go as you travel, recording "live" bands, singers, and orchestras, making a family sound-album, and suggestions for a variety of other applications. The final chapters cover such topics as picking the right kind of tape for different recording tasks, keeping your equipment in shape, and getting the best possible sound in playback.

Throughout the book, I have made an effort to avoid jargon, firmly believing that even technical matters are often best said in plain English. Where it has been necessary to employ special vocabulary to discuss specific concepts, each technical term is explained the first time it appears. Consequently, this book can be read without prior knowledge of the technical aspects. Occasionally you may forget the definition of a technical term when you encounter it later on, or, if you are a skip-reader, you may run into it unprepared. In that case, just turn to the Index, which will refer you to the explanatory passage.

Durable Pleasures

CHAPTER 1

A PERSONAL INVITATION

Just within the past few years, tape recording has grown into one of America's foremost hobbies. At the latest count—conducted by reliable market researchers—it seems to be edging photography out of first place.

In fact, there are parallels. Half a century ago, when the first modern cameras and films ushered in the era of push-button photography, countless people first realized that the fleeting images of their lives could be held fast for lasting enjoyment. Today, technically advanced but easy-to-use tape recorders lead to a similar discovery: that the fleeting *sounds* in people's lives can also be held fast to become durable pleasures.

But the pleasure you gain from your recordings depends on your skill in making them. As a general proposition, I think it's safe to say that almost anything becomes more enjoyable if you do it well—and that's the reason for this book. It is a personal invitation to increase your skill and knowledge as a recordist. As a practical guide to better recording, this book has three basic aims:

1. To show you how to get optimum results in different types of recording.
2. To help you explore a greater range and variety of things to do with tape.
3. To explain the essential performance factors by which you can judge the quality of equipment and enable you to pick the equipment best suited for your purpose.

Before tackling the details, it may be a good idea to consider our topic in a broader perspective. With the techniques of recording growing more sophisticated and imaginative, recording is beginning to be recognized, if not as an art, at least as an important creative craft that is changing the character of music in our time. The highly personal "sound" of a rock group, for example, is largely the creation of the recording engineer and his arsenal of electronic tricks and devices. The musicians could not attain the desired effect without the contribution of the recordist. In effect, modern popular music is the result of a new type of collaboration between artist and engineer. As sociologist Marshall McLuhan points out, the medium becomes the message.

Even traditional music benefits from the possibilities inherent in modern recording. Opera recordings often project the composer's musical intent more convincingly than a "live" performance, simply because the balance between singers and orchestra can be electrically controlled to surmount the limitations of the human voice. And a well engineered symphonic recording reveals more of the score than can be heard in the hazardous acoustics of many concert halls. Most important, most music heard in the world today is no longer heard in the presence of the performer but is carried to the listener through recordings. In this way, music has become independent of time and place. A performance can span the world and can be recalled in all its immediacy years after its original sound has vanished. This, it has been said, is perhaps the most significant development in the entire history of music.

Creative recording is by no means restricted to music. The impact of many films and TV productions owes much to the subtleties of the sound track, even though most viewers remain unaware of the audio methods now used to engage their emotions.

In this book, of course, we are not concerned with commercial studio recording. On the contrary; my point is that the creative techniques of modern recording are no longer confined to the professional studio. Just recently, sophisticated home recording equipment has been designed to put these advanced techniques at the disposal of the amateur, and this book will suggest how to use imaginative recording methods on your own projects.

Even if you do your recording in a more casual way, you have much to gain from recent developments. For example, cassettes have now reached a point of perfection where they offer sound quality on a par with traditional open-reel tape; and the sheer convenience, compactness, and simplicity of cassettes are rapidly making personal sound recording part of everyday life. A major section of this book will therefore ex-

Compactness, convenience, and ease of operation are the prime virtues of the cassette format. Just snap the cassette into the deck and you're ready to record or play. (*Courtesy 3M Company*)

plore the amazing versatility of the newly improved cassette medium in a wide variety of applications.

If you are a bit hazy about the exact nature of the recording process, a chapter on the theoretical underpinnings of magnetic sound recording will dispel some of the fog. Such theoretical understanding will help you choose your equipment intelligently and help you get the best possible performance from the equipment you own.

But in one respect, I hope you will retain a sense of mystery about sound recording. This has to do with the basic motive underlying all forms of recording—in pictures, in writing, or in sound. It is simply the uniquely human urge to give permanence to what is transitory. I don't mean to suggest that every time you turn on your tape recorder you are creating documents of timeless and universal value. On the contrary, it is the intimate and personal aspect of private recording—holding fast the music or the events that are important to *you*—that lends personal recordings their charm and significance. And yet, there is a larger dimension that relates each of us with our tape machines to a greater community extending far beyond the bounds of the individual. Ever since our stone-age ancestors painted their hunting scenes on the walls of their caves, mankind has tried to fortify the experience of life against the threat of oblivion. All painting and all writing are aspects of this perennial longing to "record" whatever holds for us meaning, profit, or delight. In this larger sense, the new technology of sound recording is but a new phase in the immemorial effort of all of us to mark our existence in the stream of time.

CHAPTER 2

FIRST CHOICE

If you are about to buy your first tape recorder, chances are you feel like a kid in a candy store: decisions, decisions, decisions . . .

The sheer variety of tape equipment available today can be quite bewildering. Imagine yourself walking into an audio shop asking simply for a tape recorder.

"A tape recorder? Certainly!" the clerk might answer. "Would you care for an open-reel deck? Stereo or quadraphonic? A cassette deck? With or without Dolby? An eight-track cartridge model? Something with built-in playback amplifiers? Or maybe a battery-powered portable?"

Obviously, you had better narrow down the field before confronting a salesman. It helps to keep in mind that the various types of recorders were designed for different purposes. Once you sort out the equipment in terms of its intended use, bewilderment vanishes and you can pretty well pinpoint exactly what you need to meet your own requirements.

Three Formats

You have a choice of three basic systems—or "formats," as they are called:

1. Open-reel recorders
2. Cassette recorders
3. Cartridge recorders

They all function on the same principle—by magnetizing a tape so that it carries the magnetic imprint of the signal representing the original sound. But that's where the similarity ends. In other respects, the three tape formats differ radically. In fact, they are mutually incompatible; that is, you can't play recordings made in one format on machines designed for another format. The first choice, therefore, is to pick among the three. Here, in brief, are some of their respective virtues and vices:

1. Open-Reel Recorders. That's the kind where you can see the two tape reels out in the open, spinning side by side. It is the oldest of the three formats, being a German wartime invention, and it still offers the best fidelity by a small but perceptible margin. If the utmost quality in sound is your overriding goal, an open-reel recorder is your obvious choice. This is especially true if you plan to record a lot of "live" musical performances and want to capture these musical efforts with the optimum sound attainable. Besides, only an open-reel machine lets you edit your tape after it has been recorded. It lets you cut and splice the tape to weed out musical clinkers. A lot of musicians use open-reel recordings to make several "takes" of the same piece and then splice together the best parts of each performance. Or, if you are recording rambling interviews or conferences, an open-reel machine lets you cut the tape to keep just the highlights. Open-reel tape also lends itself most readily to making a "montage" of sounds from various takes and to other more complex techniques involving the blending of various inputs (e.g., speech and background music, or adding sound on sound so you can play duets with yourself). So, if you are itching to do the kind of "creative recording" I mentioned earlier, an open-reel recorder will give you the widest range of expressive options. Finally, if you are into quadraphonics (four-channel sound) an open-reel recorder equipped for this type of operation is by far the best for the purpose. In sum, the open-reel format is the preferred choice for the serious hobbyist.

2. *Cassette Recorders.* The payoff here is simplicity, compactness, and convenience. For sheer ease of operation, you simply can't beat the cassette. Just snap it in, push the button, and you're ready to record or play. What's more, the system is virtually foolproof. Not even your most fumble-fingered friend —not to mention members of your family—can get it snarled up. Here the moving tape is wholly enclosed in the cassette. Unlike with open-reel tape, there are no loose ends dangling, nothing to thread into the recorder, and nothing to stretch, snap, spill, or otherwise mutilate. By ingenious design, all doors have been closed to possible disaster. This alone makes the cassette format ideal for beginning hobbyists and for general family use.

Fortunately, the attractive simplicity of the cassette format has been gained at only the slightest compromise in fidelity. A good cassette machine produces sound of a quality nearly equivalent to that of open-reel tape. This was not always so. When cassette recorders first appeared on the market less than a decade ago, they seemed little more than toys, fit only to take to the beach, take dictation, or immortalize baby's precious gurgles. Not even Philips of Holland, the giant electronics firm that originated the cassette concept, could have foreseen that the cassette would ever challenge open-reel tape as a serious recording medium. Actually, the development of the cassette to its present level of excellence has been quite recent. Consequently, earlier prejudices against the cassette on grounds of inadequate fidelity still persist among purists. Therefore, I should like to emphasize that a top-grade cassette recording is entirely satisfactory even to critical listeners. Only if you make recordings of a "live" performance are you likely to notice any quality margin in favor of open-reel tape as compared to cassettes. And chances are that even then you'll notice the difference only in the loudest orchestral climaxes, where a very strong and complex signal has to be captured on the tape. This high level of fidelity, plus unsurpassed ease of operation, now makes the cassette the preferred recording format for almost anyone who does not insist on tape editing.

Is the cassette then the logical choice for you? Let's put it this way: If you anticipate that your main recording activity will consist of taking radio programs off the air or copying your friends' phonograph records, and if you plan to do only rather casual "live" recordings via microphone, you will probably find a high-quality cassette machine most satisfactory for your purposes.

Cassette machines have yet another attractive feature. As a group, they represent excellent dollar value. Generally priced lower than the more sophisticated open-reel recorders, cassette models offer a surprisingly good cost/performance ratio, particularly in the $250–$400 price range.

3. *Cartridge Recorders.* It's easy to confuse cartridge and cassette, partly because the two words have similar meanings and because the two items look somewhat alike. The cassette is a thin package, hardly more than a quarter-inch thick, about the size of a small pocket notebook. The cartridge is somewhat bulkier and looks like a lopsided pack of king-size cigarettes. Yet in design and performance, there are big differences. The cartridge had been originally designed for use in car stereo systems, where fidelity is not a main requirement. After all, you can't hear the finer nuances of music amid what Cole Porter once called "the roaring traffic's boom." Despite attempts to domesticate the cartridge for home use, some of its limitations have stuck. So far, at any rate, the sound quality in this format does not quite measure up to stringent high-fidelity standards, and some engineers claim that the drawbacks are inherent in the mechanism of the cartridge itself.

Besides, cartridges are awkward to use. They have eight

parallel sound tracks on the same tape, which makes it hard to locate the particular piece of music you want to hear.

With all these drawbacks, the long-term prospects of the cartridge medium appear doubtful. There are rumors and signs that even car makers will eventually switch to the better-sounding and more convenient cassette format and gradually allow the cartridge to fade out of the picture. Certainly, as a home recording medium, the cartridge is a good thing to stay away from. At least that's my personal view.

Offhand I can think of only one situation where you might want to buy a cartridge recorder for the home. That's if you have a cartridge player in your car and want to make your own recordings at home for playing on the road. However, if you're starting from scratch, stick with cassettes all the way—both at home and in the car. You'll get better sound and greater convenience. Besides, cassettes are a lot more compact, which is important if you want to stock music for a long trip in the glove compartment. Four cassettes, holding up to eight hours of stereo music, fit in the space taken up by a single 60- or 90-minute cartridge.

So far, I have given you a quick thumbnail sketch of the three basic formats. Another major grouping in tape recorders is between battery-powered portables and equipment designed for fixed installation in the home. I'll be talking about on-location recording and portable recorders in a separate chapter. For the present, let's assume you are mainly interested in a home installation.

The Deck Concept

Some tape recorders—either open-reel or cassette—come with their own loudspeakers and with built-in amplifiers to drive those speakers in playback. But the quality of such equipment is usually only fair to middling, the obvious limitation lying in the playback speakers and amplifiers. Most qual-

Basic types of tape recorders

a. A high-quality open-reel tape deck by Teac.

b. A typical stereo cassette deck by Sansui, featuring the Dolby noise reduction system.

c. A quadraphonic recorder by Technics using eight-track cartridges.

d. A miniature portable cassette recorder by Sony, small enough to fit in the palm of the hand.

ity tape recorders are therefore built on the "deck concept."

The term "deck" means that the recording equipment does not have its own playback facilities. Rather, it plugs into your home-based stereo system, playing through your regular amplifier and loudspeakers. Thus the tape deck becomes an integral part of your home music setup. It takes advantage of the full power and range of your sound system. And since the tape deck stays permanently connected to the rest of your system, it remains ready for instant operation whenever you want to tape something off the air or copy a phonograph record. In styling and dimensions, the deck fits in with the other components of your sound system, and you can arrange it to keep the connecting cables neatly out of sight. Of course, all decks provide facilities for "live" recording. All you have to do is plug in the microphones.

Thanks to their popularity, there is a far greater choice of decks than of any other kind of high-quality stereo recording devices. Whether you decide on a cassette deck or an open-reel deck, you are likely to find one with just the features you want—and maybe even one in a price range you can afford.

Quality Checks

You can make a quick performance test of any tape recorder even without a critical understanding of its technical specifications. You don't even need elaborate equipment. The only test instruments required are your ears. This doesn't mean that such informal quality checks are unreliable. On the contrary. As any engineer will tell you, the human ear is still the most sensitive indicator of sound quality, and after they are through with their exacting technical measurements, audio engineers still rely on listening tests for the ultimate confirmation.

The test procedure for these ear-checks of tape equipment is very simple. You dub a phonograph record onto tape. Then, as you play back the tape, you switch back and forth between the tape recording you just made and the original disk. The idea is to observe just how closely the recording duplicates the source. With a good machine, you should hear virtually no difference.

Of course, only a very good phonograph record should be used as a sound source. It should have very quiet surfaces, free of scratches, and you should dust off the disk with a good record cleaner just before the test. Obviously, the record should not be worn. If it has ever been played on an ordinary garden-variety phonograph, forget it; for even a single play on an inferior turntable can permanently damage a disk, causing fuzziness and distortion in the loud passages. Also, the record should contain certain kinds of sound that really put a tape recorder on its mettle: shimmering cymbal crashes and bright trumpets to test the clarity of the highs, some very low notes to test response in the low range, and—most importantly—some long-sustained notes on the piano. The latter are a crucial test for what is known as "wow and flutter"—the tremulous unsteadiness of pitch that is the worst bugaboo of poor recorders. If the long-held note trembles and wavers, it's a sure sign that the machine you are testing is inadequate for good musical recording.

As you compare the tape copy with the original recording, listen in particular for differences in brightness and clarity of sound. Do your listening at fairly high volume level, because the ear is far more sensitive to distortion at greater sound intensity and therefore more likely to detect flaws in the equipment. Make sure you are listening to the original and the recording at the same loudness level as you switch back and forth. Otherwise, the ear is fooled and the comparison is invalid. Needless to say, the playback equipment (turntable, amplifier, and speakers) used in this test should be of very good quality so as to show up the subtle differences in sound on which these evaluations depend.

When you make your test copy from the original disk, be sure to use high-quality tape. So-called low-noise high-output tape is preferable, for both open-reel and cassette. Open-reel machines should run at a speed of 7½ inches per second for this test. (The slower alternate speed of 3¾ inches per second featured on most open-reel recorders does not allow the machine to show its best performance.) For the first test recording you make, let the recorder operate during loud passages at a safe margin below the maximum level as indicated by the recording meters. Then, as a special "torture" test, repeat the comparison with the recording level set so that the meter touches the limit during the loud parts. Then compare the two "takes" while adjusting for the difference in playback volume. If the higher-level recording does not have higher distortion and noticeably more hiss than the first take, it is a sign of good overall design of the recorder.

You can extend this test method to compare two or more separate recorders. In that case, record the same passage from the phonograph record on both the tape machines you want to compare, using the identical tape type. You can then compare the two recorders with each other by switching back and forth between them on playback. And, of course, you can compare each recorder against the original disk sound. Again, to make these comparisons meaningful, you must adjust the playback volume of all three sound sources (the two recorders and the original disk) to the same loudness level.

The listening criteria for comparing two tape recorders are the same as for comparing a tape copy to a disk original. As you listen, ask yourself specific questions to focus your attention on telltale quality indicators: Are the extreme highs free from fuzziness? Is the overall texture of orchestral sound clear and transparent? Is the deep bass clearly defined, or is it just an indistinct thud? Also check for quietness of background. There should be no hum and very little hiss.

Of course, the quality of the playback amplifier and loudspeakers also affects what you hear. But since you are using the same amplifier and speakers in all your comparisons, these factors remain constant and any differences you hear are those between the recorders.

Personally, I find a good symphonic recording the best basis for comparisons, simply because a large symphony orchestra provides the greatest variety of different sounds. This is especially true in a concerto featuring solo piano in addition to the orchestra. Besides, a massive orchestral climax is the most crucial test for overall quality in audio equipment. Still, if your main interest lies in some other kinds of music, by all means repeat these ear-checks with the type of music that means most to you. If you are an opera buff, for example, try voices. A ringing tenor, or a soprano letting go with high C will give any tape machine a real workout. If jazz turns you on, play a well-recorded combo. Watch the trumpet. Does it have its natural bright timbre? Does the bass fiddle come through with a real sense of solid lows? Does percussion sound crisp and clear so you can imagine the drumstick hitting the skins? All these are vital quality indicators. I would caution against only one kind of music as test material: rock, as a rule, makes a poor test because many of the recordings are so gimmicked that they give you no adequate reference standard in terms of the natural sound of the instruments.

Just one more caution: never try out a tape recorder by recording your own voice through the microphone. Your own voice will always sound unnatural to you on recordings. The reason is that when you speak, you hear yourself mainly by bone-conduction. The sound travels from your throat and mouth directly to your inner ear through your skull. Obviously, when you listen to a recording of your voice, this is not the case. Consequently, it will sound quite different to you and offer no valid basis for comparison.

Aside from these listening tests, you should also make some basic mechanical checks in evaluating a tape recorder. Put the machine through its paces. Work all the controls. Hit all the buttons. Make it stop and go in rapid sequence. Switch from

FAST FORWARD to REWIND and vice versa, and note the smoothness of the mechanical action. In an open-reel recorder, watch the motion of the tape. The tape should neither slacken nor pull. If it slackens, there is danger of a tape spill. If it pulls, the machine might stretch or break the tape. Either event is disastrous if it happens in the middle of a recording session. A good machine treats the tape very gently, maintaining fairly constant, regulated tape tension, no matter what you do.

If you are testing a cassette recorder and find that the tape binds or jerks, try the same recorder with another cassette. Often such difficulties are due to a faulty cassette rather than a faulty recorder.

As for the listening tests, they are largely subjective and are not meant to be a substitute for the objective performance measurements done in the laboratory. Still, they give a fair indication of overall performance. Most importantly, such tests give you a chance to discover your personal reaction to a particular recorder under typical conditions of use. In short, such tests enable you to predict your personal compatibility and satisfaction with a particular model. And that's what really counts.

Shopping Strategy

Not all dealers will let you conduct such elaborate tests, especially not the discounters who depend on quick turnover or the catch-all appliance merchants who just happen to carry a few audio items along with washing machines and toasters. But specialized audio dealers may gladly set up such a test for you in their showrooms, providing you don't come barging in on a Saturday or at the height of the Christmas rush. Many specialized audio dealers are hobbyists gone pro and may actually welcome the chance to demonstrate their merchandise at length and otherwise talk shop with you. If you are looking for such a store, a shop bearing the emblem of the Institute of High Fidelity—the trade association of the audio industry—is your best bet, and you can find the listing of such shops in the Yellow Pages in many localities. And while you're on the phone, it might be a good idea to let the dealer know in advance that you want to conduct listening tests with tape recorders. He can then get the equipment hooked up for you before you come, and that will make your trip to the audio showroom much more pleasant, relaxed, and profitable. At best, it will be a real ear-opener to help you find your way from initial confusion to rational choice.

CHAPTER 3

MONO, STEREO, AND QUAD —QUESTIONS OF AMBIENCE

My friend Sonja is four and a half years old, and her way home from kindergarten leads under a low overpass. Invariably, she will stop right under the stone arch of the bridge and let go with a blood-curdling Indian whoop. Sonja, it seems, has discovered the principle of acoustic ambience, one of the most complex and elusive aspects of sound. At any rate, she is delighted at the reverberation under the bridge.

Sonja's fascination with sonic ambience is shared by countless bathroom baritones savoring the ego-building echoes of glazed tile, by architects agonizing over the design of projected concert halls, and by recording engineers experimenting with mike placement and channel-mixing to infuse the right sonic "feel" to their latest recording. They all are intensely aware that acoustic surroundings contribute much to the character of what we hear. This sense of ambience—the feeling of "acoustic space"—greatly affects our experience of music. A poorly designed concert hall or an improperly miked recording can drain the life from an otherwise fine performance. Conversely, the right kind of ambience properly captured in a recording greatly enhances the emotional impact of the music. That is why in recent years much engineering effort has gone into refining the ways in which ambience is conveyed in recordings. Which brings us to the sometimes rather confusing topic of mono, stereo, and "quad."

Mono, stereo, and quadraphonic (usually called "quad" for

short) are methods of sound reproduction differing in the number of signal channels used to capture and transmit the sound.

Mono (short for monophonic) employs a single channel. It is the oldest form of sound reproduction and still prevalent in portable radios and portable recorders where compactness is essential and all the equipment must be contained in a single unit. To convey the basic sound of speech or music, all you need is mono. What then do stereo and quad accomplish that mono can't? Essentially, stereo and quad are attempts to convey the ambience–the acoustic space–in which the sound occurs.

The same music played by the same orchestra may sound quite different in Carnegie Hall in New York than in Symphony Hall in Boston. Many concert halls, such as Vienna's Musikverein or Amsterdam's Concertgebouw are world-famous for the splendid glow they impart to orchestral music. Multichannel recording tries to capture this sort of acoustic ambience and deliver it to the listener as a kind of bonus along with the music. Producers of popular records often create musical ambience effects through echo chambers and other electronic devices that are as much a part of the total sensory impression as the music itself. The personalized "sound" of many rock groups, as we have seen, is created mainly by such technical manipulation of acoustic space.

It is to reproduce these "space factors" in the home that the techniques of stereo and–more recently–of quad have been developed. Yet in another sense, the principle has been around a long time. Actually, stereo was "invented" millions of years ago when two-eyed, two-eared creatures first appeared on earth. We perceive the world through a two-channel system of vision and hearing. This is what makes our perception of sight and sound three-dimensional, giving us the experience of visual and acoustic space.

Just how we perceive space has never been fully explained, though experimental psychologists have been tackling this particular question for more than a century. As a rule, we are not conscious of the three-dimensionality of our perception. We just take it for granted. But if you want to give yourself a dramatic demonstration of how important our two-channel sensory equipment is to our basic orientation in the world, just try to thread a needle with one eye closed.

Something analogous holds true of stereophonic hearing—that is, hearing with two ears. Until two-channel stereo was introduced in recording and broadcasting in the late 1950s, the space factor in music could be only imperfectly suggested by these media. True, even a monophonic recording captures some of the reverberation (or lack of it) from the space surrounding the musical event. But at best, mono sound compares to the "real" three-dimensional sound as a perspective drawing compares to the real object. Since only a single monophonic channel carries the signal, all the sound seems to issue from a single spot. If you play a small portable radio or tape recorder with only one channel, you sense the spatial restriction. This doesn't matter much in reproducing speech, for the original sound source—the mouth of the person speaking—represents only a single location in space.

But for reproducing music, stereo is far preferable. With two separate recording and playback channels, stereo provides a second "ear" for the recording equipment. In short, it enables the recorder to "hear" the way you do—with three-dimensional space represented by the *difference* between what goes into either channel.

The most obvious aspect of stereo hearing is directionality. Even with your eyes closed you can tell the location of the players on the stage or in the studio when you listen at home to a stereo system. The sound doesn't appear to come either from the left speaker or the right. It appears to come from various points *between* the speakers where there is no obvious sound source. How does this happen? It is from the fact that the two speakers reproduce the same sound with different degrees of loudness. From this difference the human brain reconstructs an impression of the original sound field. Consequently, the stereo listener gains the impression of facing the stage full of musicians and can sense the location of each player. Thus the "stereo effect" makes music reproduction far more lifelike than mono and helps bring out details of the score that would have been obscured otherwise.

But the left-to-right directionality of stereo is only one aspect of acoustic ambience. Stereo also conveys a sense of depth. Each microphone used in a stereo recording captures different patterns of sound reflection because it is located in a different place from the other microphone. From this the listener gains a three-dimensional perception of the acoustic space of the original performance. Good stereo recording thus conveys a fair suggestion of sonic ambience.

Within the last few years, the techniques of capturing and creating acoustic ambience have been further refined by the introduction of four-channel, or quadraphonic sound. In four-channel sound reproduction, two speakers are placed to the left and right *behind* the listener in addition to the left and right speakers in front. Because each speaker is fed by a separate signal channel, it reproduces a different aspect of the total sound field. The listener is thus virtually surrounded with a differentiated sound presentation. It should be emphasized that this occurs only in true quad, that is, if all four speakers receive separate and different signals. Simply hooking up two extra rear speakers to a stereo set will not result in true quad ambience.

This elaborate setup raises a basic question: why do two-eared people need four-channel sound? Stereo, as we have seen, corresponds to our normal mode of hearing, the two channels corresponding to our two ears. Why then the extra rear channels in quadraphonic systems?

The reason is that in real life we hear sound all around us—not just from the front. In this sense hearing is different from vision. We can't see in back, but we hear in back. Quad is an attempt to take account of this fact in music reproduction.

Two stereo speakers, properly spaced out, give us a fine

frontal presentation of the music with locations of the players well defined. But if you are sitting in a concert hall or any other room where music is played, a large percentage of the total sound hits you from the sides and the back. This is the sound bouncing off the rear and side walls of the room. Acousticians have estimated that it represents as much as 70 percent of the total sound energy perceived by the listener. Because the reflected sound path (via wall bounce) is longer than the direct sound path directly from the player, there is a time difference between what you hear from the front and from the back. It may be just a few milliseconds, but this tiny span gives your brain additional clues to the ambience—the space in which the music is played. It is these reflections, coming from all sides and from the rear, that are captured and reproduced by a quad system.

In standard (two-channel) stereo, the reflections from the rear are blended in with the sound reaching you from the two front speakers. This can suggest but not truly represent the ambience of the performance. In four-channel systems, the rear sound actually comes to you from behind. Hence, the acoustics of the concert hall are more truthfully transposed into your living room. As you listen to quad, you get the sense that your living room is much bigger than its actual dimensions. The walls seem pushed back to simulate a big auditorium.

By adjusting the four-channel balance between front and rear you can create interesting effects. For example, by giving a little more prominence to the rear speakers, you can elicit nearly as much sound from them as you do from the front speakers. The result is that you feel yourself immersed in sound. It feels almost as if you were swimming in music, a feeling that can be tremendously exhilarating at great orchestral climaxes. And with rock groups, you feel as if you were right in the middle of what's happening, with the players and singers all around you.

Open-reel tape is the ideal medium for four-channel sound

simply because there are no technical problems in putting four separate, parallel sound tracks on standard tape. True, there are other sound sources for quad, notably four-channel records and quad tape cartridges. But, as we have already pointed out, the tape cartridge has inherent limitations of fidelity, and four-channel records are highly susceptible to wear. Besides, there still are technical problems with four-channel records and they have a comparatively high level of distortion. Open-reel tape sounds noticeably better than any other quadraphonic medium. (So far, the four-channel medium has not yet been introduced on cassettes because the narrowness of cassette tape presents technical difficulties. But it's an open secret that cassette makers are working very hard on a solution to these problems, and four-channel cassettes may soon be in the offing.)

Of course, four-channel open-reel recorders—requiring extra heads and electronic circuits for the rear channels—are considerably more expensive than standard stereo recorders. Besides, the four tracks take up twice as much space on the tape as two stereo channels. Result: playing time for a given length of tape is cut in half. This makes the four-channel medium quite expensive. Is it worth the price?

My own experience may help answer that question. I have a four-channel open-reel recorder. Like all such recorders, it can also be operated on two channels in standard stereo. I find that most of the time I do not use the four-channel facilities. For one thing, there isn't all that much music around that I want to record in four channels. True, there are quad broadcasts and the catalog of quad recordings is steadily growing, and I have occasionally taped the broadcasts off the air in four channels and also copied my friends' four-channel records. But I find that a good stereo recording (disk or tape) played over all four speakers sounds almost, if not quite, as good as recordings with four separate channels, and because of the saving in tape, the cost per minute is exactly half. This is especially true if the amplifier or receiver through which these stereo recordings are played is equipped with a so-called matrix for "local enhancement."

Even in making "live" recordings of an actual musical performance, I find two-channel stereo entirely satisfactory in most cases. For one thing, you use only two microphones instead of four, which is far easier and less cumbersome to set up. And most of the musician friends for whom I make tapes actually prefer two-channel to four-track recordings because they can be played back on most standard equipment. (As yet, very few people own four-track tape recorders.) I have used all four tracks in "live" recordings only on a few occasions when I recorded professional performances with the idea of perhaps offering the tape commercially to a record company.

When four-channel sound was introduced, the initial reaction of many audio fans was rather skeptical. To some it seemed like a brash gimmick on the part of the audio industry to sell more equipment. After all, you have to double the number of speakers and amplifier channels to "go quad." In short, it seemed that the industry was just treating itself to a 100-percent hike in hardware requirements.

Personally, I do not share this cynical view. I think that four-channel sound offers a true advance in sonic realism. However, the net improvement is comparatively slight. Compared to a good stereo recording, quad offers superior ambience. But I am not convinced that the difference is great enough to justify the added cost in either complexity or cash.

Whether you go quad or stick with stereo is ultimately a purely personal choice. Only you can decide whether the difference matters enough *to you* to pay the price and give houseroom to those extra speakers.

There is a simple way to help yourself with this decision. Play a four-channel tape at your audio dealer's demonstration room. Then, as you listen, switch to standard stereo and cut out the rear speakers. The difference is then clearly spelled out for your own ears. But to make this side-by-side comparison valid, make sure that the ambience information contained on the rear channels is not entirely cut out as you switch to stereo. It should be blended with the two front channels and appear in the front speakers. The switching arrangement for doing this differs on various models.

Of course, if you buy quad equipment, you have the best of both worlds. You always have the option of running it in stereo—just two channels at a time, or even in mono, using each of the four channels separately. Similarly, you can also operate a stereo recorder in mono. The obverse is not true. You can't get quad or stereo from a mono recorder or quad from a stereo recorder. In other words, you can operate a recorder on less than the maximum number of channels it is built for, but you can't get extra tracks if there is no provision for them.

To sum up, let me suggest a few basic guidelines for choosing between mono, stereo, and quad:

Mono: If you plan to use your recorder chiefly for speech, interviews, or similar applications, mono may be all you need. In portable use for on-location recording, mono recorders have the advantage of being light and compact, and many of them are designed for both plug-in or battery operation.

Stereo: Two-channel sound is the current standard for musical recordings. A great variety of equipment is available in different price ranges, in both open-reel and cassette format. On the whole, stereo gives excellent musical results at moderate cost. Microphone techniques for live stereo recording are not complicated, and setting up for a session is relatively simple.

Quad: Equipment is expensive and procedures complicated. But if you plan to make professional or semiprofessional recordings, or want to experiment with all kinds of multitrack tricks, quad offers the greatest range of creative possibilities.

CHAPTER 4

HOW IT WORKS

Our main object, up to this point, has been to scan the whole spectrum of available tape equipment to help you narrow down your particular choice from all the possibilities. By now you probably have a pretty firm notion of just what kind of recorder you want to buy. The next step is to make sure you get a really good one. What is it, then, that makes a tape recorder good?

The answer to that question will be spelled out in all its specifics in the next chapter. But first we must explore just how a tape recorder works.

In principle all tape recorders are alike. As the tape rolls through the machine it passes over a so-called recording head and is magnetized by contact with the head. This recording head—a ring of laminated metal with a small gap in front where it touches the tape—is energized by the electrical signal representing the sound to be recorded. As a result, the magnetic field that develops across the narrow gap in the recording head varies exactly in proportion to the sound waves originally captured by the microphone. As the tape glides across the gap in the recording head, a magnetic replica of the sound wave is thus imprinted on it.

In playback the process is reversed. Again there is interaction between a magnetic head and the moving tape. But this time the magnetic pulses carried by the magnetized tape induce an electric signal in the head as the tape travels past

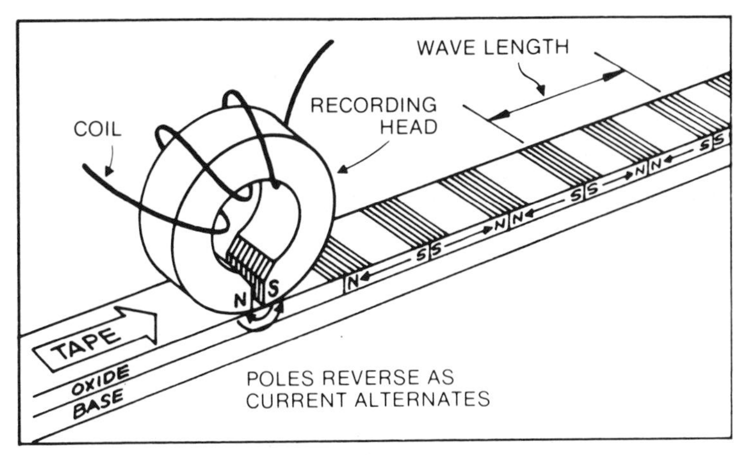

The basic principle of magnetic recording. The electric pulses from the coil magnetize the recording head, with magnetic poles reversing as the signal current alternates. The magnetic traces of the different audio frequencies (wave lengths) are then imparted to the oxide layer of the tape as it moves past the recording head. (*Courtesy 3M Company*)

A highly magnified photograph of a recorded track on tape. The tiny iron particles on the tape are grouped by the magnetic impulses to represent the sound waves of the music. (*Courtesy 3M Company*)

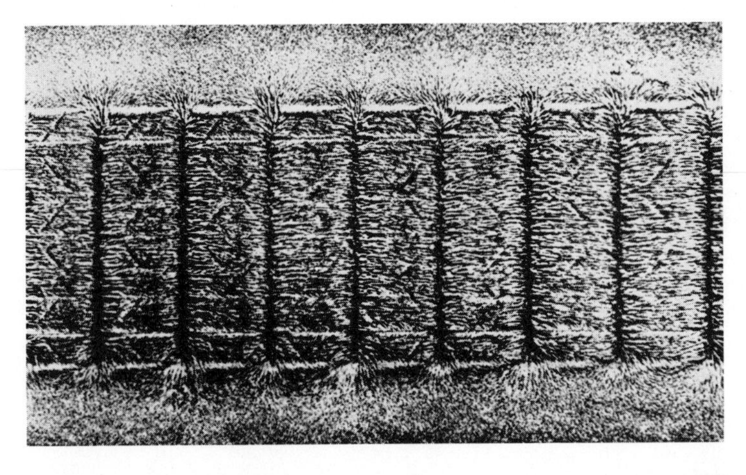

the gap. The signal so generated is then amplified and reproduced through the loudspeakers.

Double- and Triple-Head Recorders

In most of the less expensive recorders, the same magnetic head is used for both recording and playback. Depending on how the head is switched into the electronic circuits of the recorder, it can either imprint a signal onto the tape or "read out" the signal from a tape already recorded. In addition, the recorder has still another head, which demagnetizes the tape before it reaches the recording head. That way, the tape is wiped clean of any prior magnetization and anything previously recorded on the tape is automatically erased as you make the new recording. This allows you to use the same tape

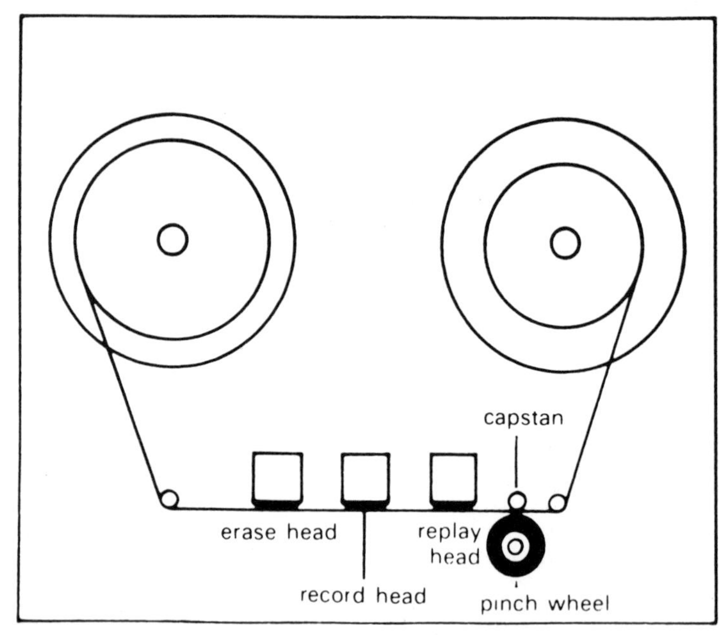

Schematic representation of basic tape recorder parts. (See explanation in text.) (*Courtesy 3M Company*)

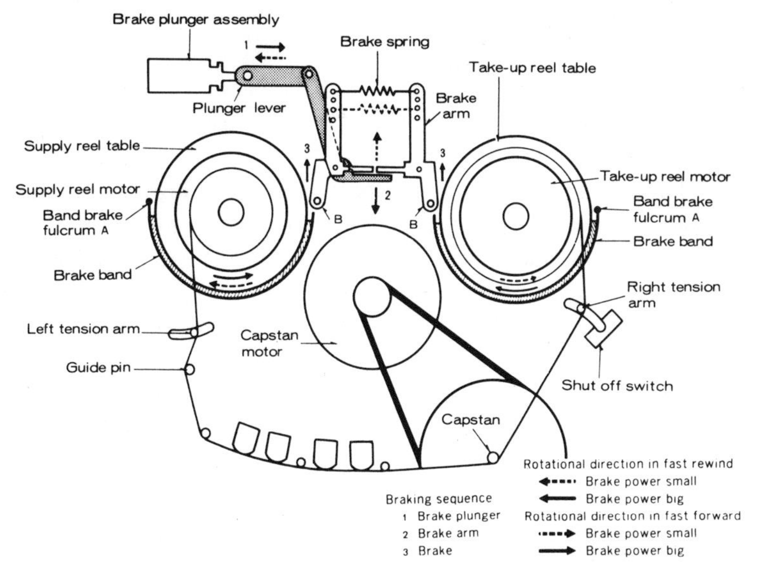

An engineering drawing of the actual working parts in a tape transport mechanism of an open-reel recorder, showing the finely adjusted braking system that assures smooth starting and stopping while maintaining proper tape tension. (*Courtesy Technics*)

The tape is pulled past the magnetic heads by being squeezed between the rotating capstan and a rubber-rimmed pressure roller, also called a pinch-wheel. Absolute roundness and precise vertical alignment of the rotating parts is essential to smooth tape flow. (*Courtesy Teac*)

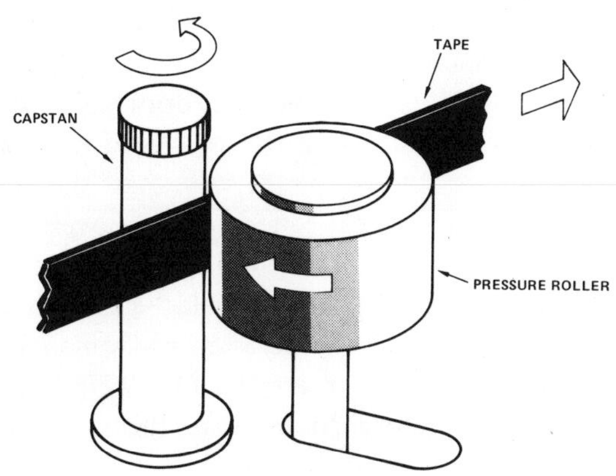

over and over again, assuming you don't want to keep whatever you had recorded before.

This two-head configuration—combined record/playback head plus erase head—is the most common pattern. Yet on the more elaborate recorders, separate heads are used for recording and playback. Along with the erase head, this makes a total of three heads. This has the advantage that both the separate record and playback heads can be optimized for their particular functions, usually resulting in superior frequency response in playback. Moreover, the separate playback head can be used to scan the tape while it is being recorded by the record head. In this way, you can listen to or "monitor" your recording while you are making it. You then know right away how the recording is turning out and can immediately make the necessary control adjustments if it doesn't sound right. If you are recording an unrepeatable performance or event, knowing you will never get a second chance to correct possible goofs, a three-head recorder with such instant monitor facilities can be a great reassurance.

Head Quality

The quality of these heads is a crucial factor in tape-recorder performance. The narrower the gap, the greater the frequency range that can be recorded or reproduced. You might compare this gap to the point of a pencil drawing lines on paper. The finer the point, the more detail can be drawn in a given space. The "paper" in our case is the tape, the "drawing" is the magnetic pattern to be imprinted on the tape, and the space for the drawing is the length of tape traveling past the head in a given time span. Since musical sounds contain overtones with frequencies up to 20,000 Hz (Hz, or Hertz = cycles per second), up to 20,000 magnetic pulses should be imparted to the tape every second. Some of the best recorders manage to do this, though a frequency response up to 15,000 Hz is generally considered quite satisfactory. In any case, this means that the magnetic marks must be very narrow so that all the pulses can be crowded together into the available space on the tape. Hence the need for a narrow gap, for the fineness of the magnetic imprint is limited by the gap width—just as the fineness of a pencil line is limited by how well the pencil has been sharpened. On high-quality machines, the head gap may be as narrow as one ten-thousandth of an inch.

To produce such heads requires precision manufacture, for the edges of the gap must be absolutely straight and smooth, and permissible tolerances are minimal. Moreover, the metal laminations from which the head is assembled must be enormously hard so as to prevent the constant abrasion of the head by the moving tape from wearing the edges down and thus widening the gap. Most recorders employ special alloys known as ferrites, which are highly wear-resistant but difficult to shape to such exact dimensions. The extreme care and rigid quality control required to produce first-rate heads add considerably

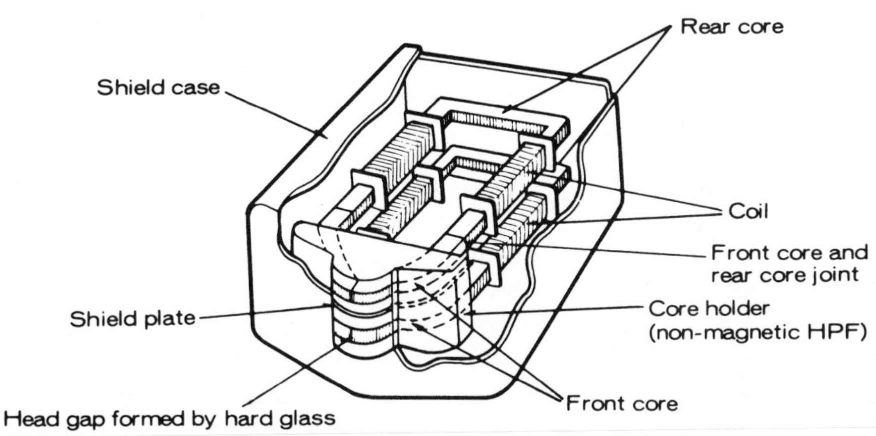

Cutaway view of a recording head showing the two coil systems and magnetic pole pieces to accommodate signals from the two separate stereo channels. In this design, the narrow gap is strengthened against abrasion from the moving tape by a glass coating over the magnetic elements. (*Courtesy Technics*)

to the cost, and this is one reason why high-quality recorders tend to be rather expensive. Moreover, even the best heads cannot function properly unless they are very accurately aligned with the tape. This calls for painstaking assembly and adjustment of the head in relation to the parts guiding the tape in its travel across the head. These are some of the hidden but vital quality factors that distinguish a really fine recorder from lesser models. To assure such dimensional accuracy over long periods of operation, high-quality recorders employ only carefully machined parts where cheap models may use stamped parts, which are inherently less accurate in their dimensions and usually more flimsy. Again, the difference shows up in price as well as in performance and durability.

You may have noticed that I place much emphasis on mechanical factors, such as precision machining and assembly of parts. The reason is that a tape recorder is primarily a mechanical device—about 70 percent mechanical and 30 percent electronic. And regardless of how good the electronics may be, if the mechanical aspects of the machine fall short of the required standards of precision, the results won't be satisfactory.

Tape Transports

One of the most important aspects of a tape recorder is the tape transport—the mechanism that moves the tape. The trick here is to maintain steady movement at constant speed. If the speed varies, the pitch of the recorded or reproduced sound varies also. The technical name for such speed variations is "wow," but in this case the term does not by any means denote delight. On the contrary, wow makes the music sound out of tune. At its worst, wow makes the pitch slide up and down like a police siren.

Another even more insidious form of speed variation is called "flutter," which occurs when the tape is not flowing smoothly across the head but moves in a series of tiny jerks and chugs. The effect of such flutter is not perceived as pitch variation but rather as a kind of tremolo or warble. It is especially noticeable on instruments where the original sound tends to be steady and devoid of tremolo, such as the piano or the pipe organ. Recording a piano passage with long-held notes and then playing it back is therefore a good test for flutter.

Some degree of flutter is unavoidable, since it is caused by the vibration inherent in all kinds of rotating machinery. The object is to keep flutter to a minimum so that it falls below the threshold of notice. To accomplish this, the drive motor must be of very high quality with all its turning parts accurately balanced and centered so that it rotates with a minimum of vibration; again, careful design and precision manufacture are the key. Moreover, any residual speed instability of the motor is smoothed out by a properly balanced heavy flywheel. Usually, drive motors of this type are found only in full-sized equipment. The lightweight motors in portable recorders tend to suffer from a fairly high degree of flutter, though the amount of wow is controlled in the better models by electronic speed-compensating circuits. A certain measure of flutter, such as is found in many portable machines, may not matter very much in recording speech, but the effect can be quite unpleasant with music, particularly in compositions containing many long-held notes.

The motor pulls the tape by means of a capstan, which is a rotating metal shaft that squeezes the tape against a rubber roller or idler, or pinch-wheel as it is sometimes called. The tape is then pulled along between the capstan and the idler as a piece of laundry is pulled through the wringer of an old-fashioned washing machine. Of course, to avoid flutter—i.e., to keep tape motion constant—both the capstan and the idler must be carefully contoured and centered and be free of surface irregularities. In good equipment, the idler is made from high-polished steel and centered on the motor shaft with extremely close tolerances.

In addition to pulling the tape past the recording and playback heads at the correct speed, the transport mechanism must also turn the tape reels. The reel which feeds out the tape is called the supply reel, and the one which winds up the tape at the end of its journey through the recorder is known as the takeup reel. The rotation of both the takeup reel and the supply reel must be properly regulated so as to keep the tape under a certain amount of tension at all times. Otherwise there would be danger of the tape spilling (if the tension is too loose) or breaking (if the tension is to tight). A combination of drag-brake and slip-clutch regulate the tension on the reels as the tape winds and unwinds, and on some of the better machines there is an additional tension equalizer in the form of a spring-loaded arm that swings back and forth to compensate for any momentary looseness in the tape.

In addition to its regular motion during recording and playback, the transport must also provide means for the fast-forward and rewind motions. During these modes of operation (fast-forward or rewind), the capstan is removed from the tape and the tape is lifted from the heads so that it can move swiftly from reel to reel in either direction. For stopping during rewind or fast-forward motion, there are brakes which must be carefully adjusted so that the tape is neither stretched nor allowed to spill loose during sudden stop-and-go changes. The smoothness and reliability of these mechanical operations are the mark of a good machine.

Many tape recorders use a single motor to drive both the capstan and the reels, the latter being linked to the motor by a system of belts and pulleys. These belt linkages occasionally require adjustment after prolonged use. That is one reason why the better machines often have an extra motor for driving the tape reels, separate from the motor that turns the capstan. The most expensive recorders use three separate motors, one for the capstan, one for driving the reels during recording and playback, and one motor devoted exclusively to the high-speed operations of rewind and fast-forward. In this way, each motor can be optimized for its particular task. Consequently, three-motor recorders offer particularly fast rewind. Also, this arrangement permits all the various kinds of tape motions to be controlled through electrical relay-switches that respond to the slightest touch of a control button. It is a pleasure to have a machine that obeys so readily to a mere touch, but this is a fairly expensive luxury; other recorders, which use mechanical control levers in place of the relay-switches, often perform as reliably and as well.

Recorder Electronics

As for the electronic part of the tape recorder, the requirement is identical to that of any other electronic audio equipment: adequate frequency range and low distortion. Audio circuits have been perfected to such a degree in this age of solid-state electronics that this is no longer a problem on any of the better open-reel or cassette decks, though in many portable recorders, circuit shortcomings still limit performance, particularly in regard to distortion.

A unique aspect of recorder electronics is the bias generator, which serves a dual function. First, it provides the erase signal which wipes out any prior recording on the tape when the machine operates in the "record" mode. Naturally, the erase head is automatically switched off when the recorder is used for playback. After all, you don't want to wipe your tapes clean as you play them.

The second function of the bias generator relates to the magnetic characteristics of the tape itself. In order to accept the signal to be recorded, the tape has to be primed by a so-called bias signal, a high-frequency signal far above the audible range, anywhere from 50,000 to 200,000 Hz. This signal is applied to the tape along with the audio signal and—for reasons not yet fully understood even in theory—makes the tape more responsive. The optimum frequency and strength of this bias depends on the kind of magnetic coating used on the tape. That is why many recorders have a switch for selecting the proper bias for different tape types.

Speed and Fidelity

The speed at which the tape moves across the recording and playback heads bears a direct relation to the fidelity attained. If you have an open-reel recorder, it is important to understand this relationship so that you can make an intelligent and appropriate speed selection. With cassette or cartridge recorders you need not worry about this, for the speed of both formats is standardized. All cassette recorders run at a tape speed of 1⅞ inches per second (ips). All cartridge machines run at 3¾ ips. Yet on open-reel recorders you have a choice of at least two tape speeds, usually 7½ ips and 3¾ ips. A speed selector on the control panel lets you pick either speed. Some reel decks also offer the slower speed of 1⅞ ips, and some professional machines feature an extra-fast option of 15 ips. The question thus arises: Why this multiplicity of speeds—and which is best for what?

One way to understand the effect of tape speed on fidelity is to revert to our analogy of the tape as a writing surface on which sonic information is "written" with a magnetic "pencil." The more surface you have available, the more information can be recorded. Suppose you're running your machine at 7½ inches per second. In that case you have a 7½-inch strip of "writing surface" on which to put down magnetically whatever sonic events occur in the music during that particular second. Let's say that back there in the percussion section of the orchestra a pair of cymbals are banged together with a resounding crash at the peak of a musical climax. Lots of shimmering highs are shooting out into the air. What's more, it's LOUD. On your 7½-inch tape strip, you have plenty of space to "mark down" all these sonic happenings. There's room enough for 20,000 separate magnetic marks representing the 20,000 Hz (or cycles per second) that spell out the silvery gloss of that ringing cymbal sound. Also, the 7½-inch tape strip contains enough tiny magnetic particles to absorb the sheer loudness of the crash, i.e., the amount of energy generated by that *fortissimo*. Net result: a clear and undistorted recording of a climactic moment that will lift you right out of your chair when you listen to it.

Now suppose exactly the same moment of music is recorded at the slower tape speed of 3¾ inches per second. Now the tape strip rolling past the recording head during that crucial second is only half as long as at 7½ ips. This means that the magnetic traces of the musical waveforms must be crowded closer together. As a result, the uppermost frequencies may be pushed so close to each other that they become at least partly undistinguishable and, consequently, the sheen of those shimmering highs may be slightly dulled in playback. The difference may be slight—so slight that it may become apparent only when you listen carefully. But a certain feeling of ease and openness in the sound is likely to be lost, especially in the type of music where a full orchestra in heavily scored passages produces very complex waveforms with lots of high-frequency content.

In regard to dynamic range—the span from the softest to the loudest sound that can be recorded—the slower tape speed also imposes limitations. With only 3¾ instead of 7½ ips, there are only half as many metal particles to be magnetized and, consequently, only half the amount of magnetic force can be imprinted on them. This means that the loudest passages have to be held down somewhat in the recording. Otherwise the tape might get overloaded and the sound become blurred at the loudness peaks. These moments of distortion may be only very brief, occurring only at the instant of high energy concentration as when the drumstick hits the skin of the kettledrum or the pianist comes down full force on the keyboard. Yet there is an unconscious psychological response to the momentary blur: for the listener the music loses something of its emotional impact.

The extra margin of frequency response and dynamic range obtained at higher tape speed is called "headroom," a term graphically suggesting that the music won't bump against any technical limits. Subjectively, this headroom lends a feeling of effortless ease and openness to the sound, especially in complex passages. As one studio engineer whimsically expressed it, "It's like having more money in the bank than you really need. You never feel the pinch, even when the dentist's bill comes in."

Recording speed is not the only factor that provides this headroom. The characteristics of the tape itself also play an important role. Some tapes, called high-energy tapes or ultra-dynamic tapes, have more magnetic particles crammed into every inch of space than ordinary tapes. Hence they can furnish almost as much headroom at slower speed as ordinary, less expensive, tapes provide at higher speed. And since the slower speed (3¾ ips) uses only half as much tape for a given time of recording than 7½ ips, it may actually prove more economical to use the better tape. Also, a high-quality playback head with a very narrow gap obtains better frequency response at any speed, which is why the better tape decks often yield very good results even at the slower speed.

All these factors must be taken into account by the expert recordist in choosing the proper speed for a particular recording job on an open-reel recorder. Based on my own experience, I might suggest the following guidelines:

I use the 7½ ips speed mainly for original "live" recording of music with a pair of good microphones capable of capturing the full frequency range. Occasionally I also use 7½ ips for copying onto tape exceptionally fine phonograph records that contain a broad frequency spectrum and a wide dynamic range. For less outstanding records, I find the 3¾ ips speed entirely satisfactory, having found that even at that speed I can usually capture the musical content of a disk without perceptible loss in transfer.

I also use 3¾ ips for taping programs from FM broadcasts, which can also be done without loss of any part of the broadcast signal since FM broadcasts in the United States are limited by law to a top audio frequency of 15,000 Hz. A good recorder loaded with high-quality tape can easily accommodate this at 3¾ ips. Moreover, most FM stations also limit their dynamic range by means of so-called compressor circuits, which automatically soften loudness peaks to avoid overloading the transmitter. In taping such broadcasts, there is little sense in expending extra tape at higher speed trying to capture what isn't there in the first place.

For recording speech, I usually find the slow 1⅞ ips speed quite adequate, particularly if the recording is to be used only for reference purposes. In taping theater presentations or dramatic readings, where subtleties of voice and diction contribute importantly to the expressive content, I find that I can capture more elusive nuances of vocal timbre at 3¾ ips.

In general, then, the slow speed of 1⅞ ips should be used only where fidelity is not a prime consideration. Therefore, if the open-reel recorder you are considering does not record at this speed, it would not be a major drawback unless you plan to do a lot of speeches or conferences. At the opposite end of the speed range, the 15 ips speed found on some professional-type recorders has also only limited applications, mainly for studio purposes where ease of editing is required. For the amateur recordist preferring an open-reel machine, a two-speed recorder designed to operate at 7½ and 3¾ ips will prove satisfactory in almost any situation.

Cassette Recorder Speed

Having examined the dependence of fidelity on tape speed, one question naturally arises: How is it that cassette recorders, operating at the extremely slow speed of 1⅞ ips, can attain adequate performance levels?

That's a loaded question. The sober fact is that most cassette recorders do not attain performance levels that could be considered even remotely adequate. The small portables usually sold at appliance stores and miscellaneous discount emporia, as well as the cassette recorders built into portable radios or cheap "package" stereo systems, generally fall far short of musical competence. Their puny drive motors churn and wobble, causing flutter and wow sufficient to make critical listeners seasick. Their electronic circuits are notoriously skimpy, smearing every sound with distortion, and their cheaply made recording/playback heads would make the frequency response of an old telephone sound pretty good in comparison. There are, admittedly, lucky exceptions to this sorry rule, and in a later chapter we shall point to some excellent standouts among portables. But the common garden variety of such recorders, while reasonably adequate for casual recording of speech, are useless for more ambitious recording projects.

By contrast, high-quality cassette decks are equal to almost any recording task and indeed rival the performance of a good reel recorder. In this case, the question of how such good quality is attained at slow tape speed becomes relevant.

The answer lies in the optimization of all the elements involved, the recorder as well as the tape. To begin with, the better cassette decks have carefully regulated precision motors with heavy flywheels that reduce wow and flutter to the point where even a long-held piano chord sounds rock-steady. And to extend frequency response at slow speed, good cassette machines employ narrow-gap heads produced and inspected with special care. Moreover, frequency response is also aided by recently developed tapes especially designed for use in cassettes. Their magnetic granules are particularly small and numerous, and very uniformly distributed over the area they cover. Thus they can register very complex magnetic patterns in a very small space, thereby largely offsetting the drawback of the slow speed. Technically, these are sometimes called "high-density" tapes, meaning that a great deal of sonic information can be densely packed into small areas of magnetic material. Some of these tapes use chromium dioxide particles as their basic material instead of the more commonly used iron oxide. These tapes now make it possible for cassette recorders to attain a frequency response to 15,000 Hz, which is ample for almost every purpose except for recording original master tapes. On some of the best available cassette decks, response extends even further toward 20,000 Hz, which satisfies even the most stringent recording requirements. Granted, that's pushing the limits of the medium and presupposes that everything is working in top form and that the equipment is properly maintained. But it can be done.

Of course, the cassette medium presents special problems in the area of dynamic range. Since the tape used in cassettes, in the interest of compactness, is only half as wide as the tape used on open reels, each track on the tape has to be much narrower. Consequently, there is less magnetic material passing by per unit of time and it is more difficult to magnetically absorb the energy levels represented in a thundering orchestral passage or the ringing note of a full-throated singer. In short, as far as loudness range is concerned, the cassette has inherently less headroom than open-reel tape. The problem is aggravated by the fact that there is also less dynamic "bottom room." By this I mean that the softness of sounds that can be recorded on cassettes is also limited. Normally, the lowest possible recording level is determined by so-called tape hiss—the hissing noise inherent in all kinds of tape. If the signal gets too soft, it may be overshadowed by this background noise, which even in the very best tapes cannot be altogether eliminated. The effective dynamic range—the range between the softest and the loudest sounds that can be recorded—is therefore restricted at both ends. Fortunately, an ingenious method was invented by which these drawbacks have been largely overcome. This is known as the Dolby, so named after its American inventor, and it is largely responsible for the break-through in the fidelity of cassette recordings.

The Dolby

The Dolby is now an integral part of nearly all the better cassette decks. It is therefore necessary to explain its function in some detail, even though—at least at first reading—it may seem somewhat complicated.

The type of Dolby employed in home-recording equipment acts as an automatic monitor that "watches" the high frequencies in the signal to be recorded. If the highs get too soft and fall below a certain level, the Dolby automatically boosts that part of the frequency spectrum. The reason for concentrating on the highs is that the hiss of the tape primarily obscures the upper frequencies while leaving the others relatively unaffected. Later, during playback, the action of the Dolby is exactly reversed. It cuts back the highs to their original level so they sound no louder than is natural. But by doing so, the Dolby also cuts back the tape hiss. Since the hiss hadn't been boosted in the first place (only the incoming signal was boosted) the hiss in playback is proportionately much less prominent. The music now stands out clearly against a background of virtual silence. By permitting softer sounds to be recorded, the Dolby extends the possible dynamic range between soft and loud. By creating extra "bottom room" at the soft end, the Dolby, in effect, makes extra headroom at the loud end. In other words, the entire dynamic spectrum is shifted downward so that it can be accommodated on the narrow cassette tape. (Of course, just because you record at a lower level doesn't mean that the recording has to sound softer in playback. The amount of actual playback volume is controlled by the playback volume setting, so—given sufficient amplifier power—even a tape recorded at low level can play back at full volume.)

Since the corrective action of the Dolby takes place only in that part of the frequency spectrum where tape hiss becomes a problem, the rest of the frequency range passes undisturbed. This selective action of the Dolby accounts for the fact that it operates entirely without musically objectionable side effects. Thus, thanks to the Dolby, the inherent drawbacks of cassettes have been effectively surmounted and no longer present an obstacle to enjoyable listening to any kind of music.

It is evident from this explanation of the Dolby principle that the process works properly only when it is used both in recording *and* in playback. A tape recorded without a Dolby will sound dull and drab when played through a Dolby circuit because its highs will be cut back without having first been boosted. Conversely, a tape recorded on a Dolby will sound shrill when played on a deck without Dolby for the opposite reason: the highs have been boosted in recording without being cut back proportionally in playback. To make sure that you can play both Dolby tapes and non-Dolby tapes on your deck, virtually all machines with built-in Dolby have a switch by which the Dolby circuit can be cut in or out, as required. (If you play a Dolby tape on a non-Dolby machine, you can usually counteract the shrillness simply by turning down the treble control. But this is merely a makeshift method for taking the hard edge off the sound and does not provide an accurately balanced frequency response. For that, you need a regular Dolby in your playback equipment.)

If the Dolby sounds rather forbidding in theory, it's easy enough to operate in practice. Just leave the Dolby switch in the ON position for both recording and playback. The machine will do the rest all by itself and you'll get cassette recording that rivals the fidelity of open-reel tapes.

CHAPTER 5

WHAT THE SPECS TELL YOU

In the preceding chapter I outlined the operating principles of tape recorders along with some general factors affecting the quality of their performance. We can now turn to the more specific performance factors—the technical specifications which provide the best available yardstick for comparing the capabilities of one machine with those of another.

When buying an open-reel recorder, the first specification to consider is the number of tracks. Catalog descriptions and salesmen speak rather confusingly of two-track, four-track, quarter-track, half-track, or even full-track models, often leaving the neophyte shopper in a bewildering haze of numbers. At a recent audio show, for example, I overheard a visitor innocently ask if a four-track recorder is twice as good as a two-track machine. I suppressed an impulse to snicker. After all, on the face of it, the question is altogether plausible.

The fact, of course, is that the number of tracks on a tape recorder has nothing to do with the quality of its performance. It merely tells how many separate recordings the machine can put side by side on a standard-width (¼-inch) tape.

To illustrate, let us take a two-track recorder. (The same machine is sometimes called a half-track recorder, the terms two-track and half-track being used interchangeably.) With such a recorder, you first record one edge of the tape from end to end of the reel. That's track No. 1. Then you flip the reel over and record the other edge of the same tape. That's track No. 2. You now have two monophonic tracks, each taking up a little less than half the width of the tape. The same two-track machine usually can also operate in stereo. In that case, both tracks are recorded simultaneously in the same direction—no reel-flipping. One track then provides the left channel while the other furnishes the right. Many professional tape recorders are of this type.

For ordinary home use, however, a four-track recorder offers important advantages. You may wonder why a machine has four tracks when stereo requires only two. The answer is that you use only two tracks at a time for stereo. But having space for two more tracks after you finish one run-through of the reel, you can flip the reel and then make an additional two-track stereo recording in the other direction. In other words, an ordinary four-track recorder doesn't really record on four tracks. It merely offers you the chance to make two two-track stereo recordings on the same tape. That doubles the total stereo recording time you can put on a given length of tape and is therefore an important economy factor. That's why the four-track stereo recorder is the preferred option for most hobbyists.

Another version of four-track recorders is adapted for quadraphonic recording, which records all four tracks at the same time, using up all the space on the tape at a single pass. Of course, if you record all four tracks simultaneously for four-channel presentation, you've got no room left for flipping the tape over for additional recording.

Four-track quadraphonic recorders give you the option of using them as standard stereo machines. In that case, only two tracks are recorded simultaneously while the other two are reserved for flip-side recording.

In short, quadraphonic recorders give you the best of both worlds—a choice of either stereo *or* quad. But there are two things to consider in their disfavor. 1. Quad recorders are far more expensive than four-track stereo machines. 2. If you are a typical amateur, unconcerned with making professional tapes, you may have little occasion to use their quad capabil-

Tape track patterns now in general use.

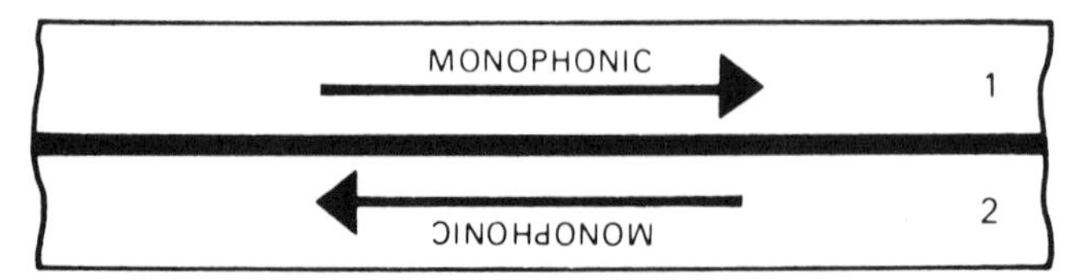

a. Half-track mono

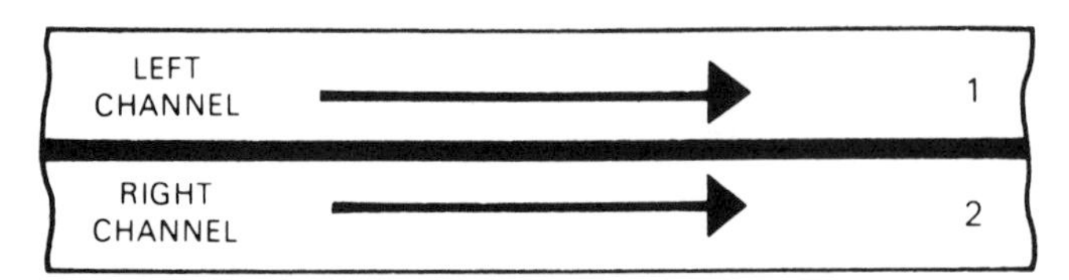

b. Two-track stereo

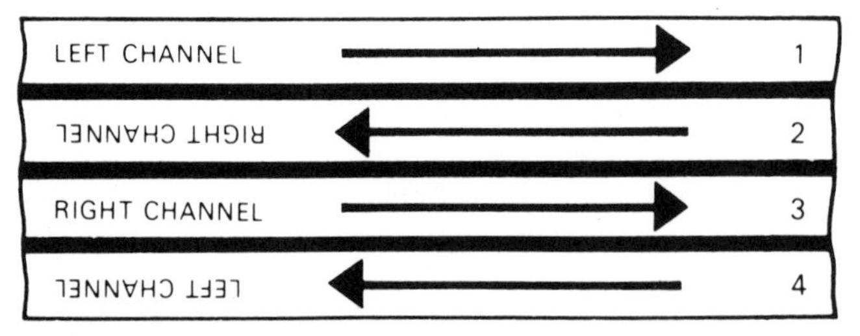

c. Four-track stereo

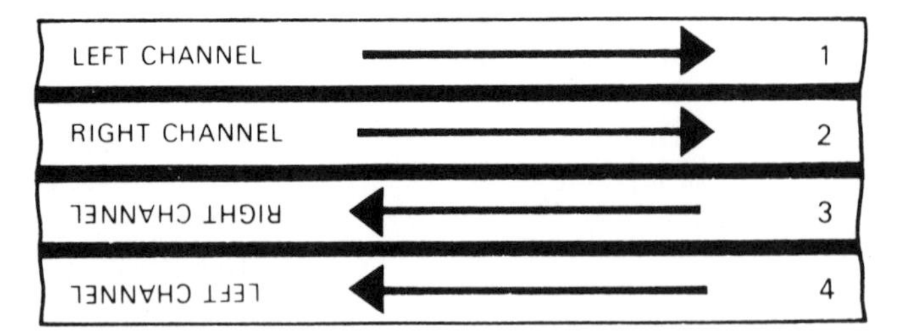

d. Cassette stereo

(*Courtesy 3M Company*)

ities. Thus, all things considered, the four-track stereo recorder still remains the best bet among open-reel machines for most hobbyists.

Frequency Response

Of all audio specifications, frequency response is perhaps the one most often cited but also most widely misunderstood. The spectrum of musical sound extends from the lowest notes at about 30 Hz to approximately 20,000 Hz. Actually, the highest pitch of any musical instrument is only about 5,000 Hz. But each note also contains a whole series of so-called overtones, which lend each instrument and each voice its characteristic tone color or *timbre*. To capture these overtones adequately, the frequency response of a good recorder should extend to at least 12,000 Hz and preferably to 15,000 Hz. If the response goes even higher, it will provide an extra margin of realism.

Often you will see the frequency response of a tape recorder stated as, for example, 30 to 15,000 Hz. Such a statement seems impressive, as it covers just about the whole required range. In fact, it is almost meaningless. Strictly interpreted, it tells you nothing more than the top and bottom notes the tape recorder can handle. What is far more important is what the recorder does with all the notes in between. What really matters is the uniformity of response between those upper and lower limits. The recorder should reproduce all the notes in precisely the same proportion as they are heard in the original music. Low, middle, and high sounds should all keep their true relation to one another. The equipment must not emphasize some frequencies more than others, nor make some frequencies weaker than others. Every part of the frequency spectrum must be given precisely its due value if the sound is to be natural and lifelike. That is what is meant by the often-heard expression "flat frequency response."

Perfectly flat response—like perfection in many other things—

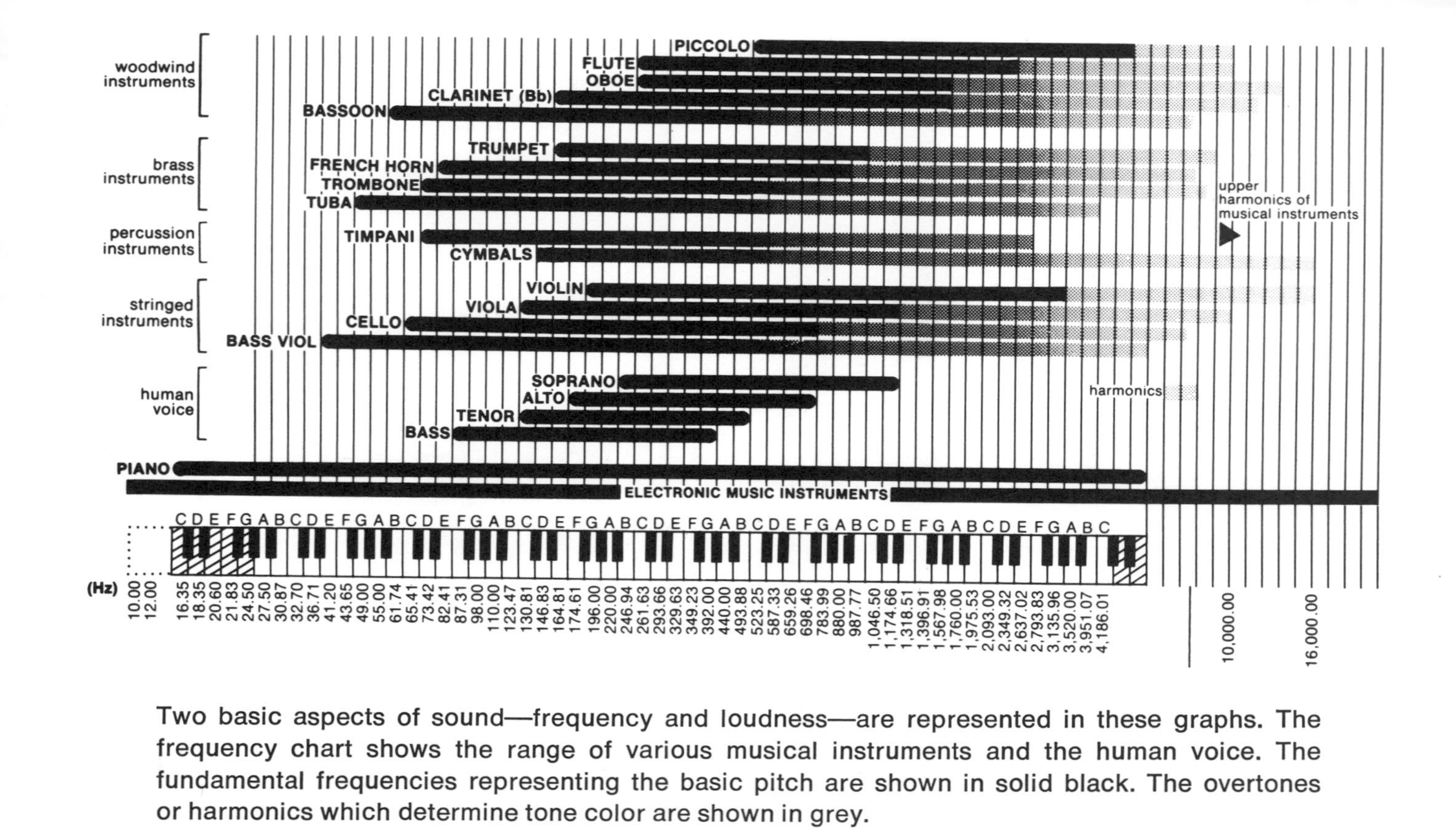

Two basic aspects of sound—frequency and loudness—are represented in these graphs. The frequency chart shows the range of various musical instruments and the human voice. The fundamental frequencies representing the basic pitch are shown in solid black. The overtones or harmonics which determine tone color are shown in grey.

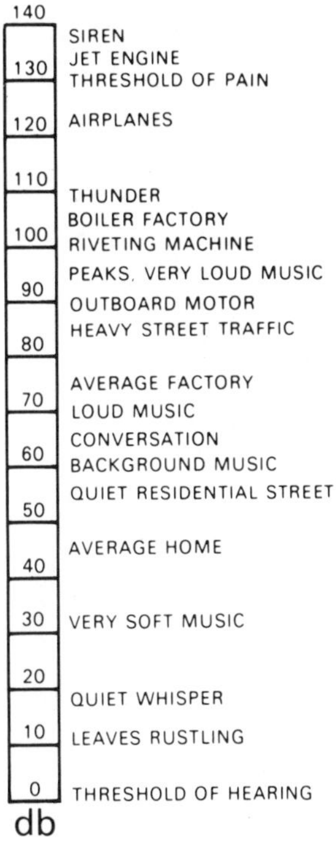

140
SIREN
130 JET ENGINE
THRESHOLD OF PAIN
120 AIRPLANES
110
THUNDER
BOILER FACTORY
100 RIVETING MACHINE
PEAKS, VERY LOUD MUSIC
90
OUTBOARD MOTOR
HEAVY STREET TRAFFIC
80
AVERAGE FACTORY
70
LOUD MUSIC
CONVERSATION
60
BACKGROUND MUSIC
QUIET RESIDENTIAL STREET
50
AVERAGE HOME
40
30 VERY SOFT MUSIC
20
QUIET WHISPER
10 LEAVES RUSTLING
0 THRESHOLD OF HEARING
db

The decibel scale is a relative measure of loudness, and the chart shows the decibel value of some typical sounds.

remains an abstraction and a perennial pipe dream of audio designers. Still, many good recorders come reasonably close to it. To tell you just how close, look at the "plus-minus" figure (prefixed by the sign ±) which should always follow the statement of the overall frequency response. For instance, the response of a tape recorder might be specified at 30 to 16,000 Hz ± 3 db. This means that at no point in the specified range does the response deviate by more than 3 db from the ideal flat response. The db stands for decibel, which is the standard measure for loudness. It is a relative measure, and the best way to interpret it is to remember that 1 or 2 db is about the smallest loudness difference detectable by the human ear.

On a good recorder, this deviation should not exceed ± 3 db,

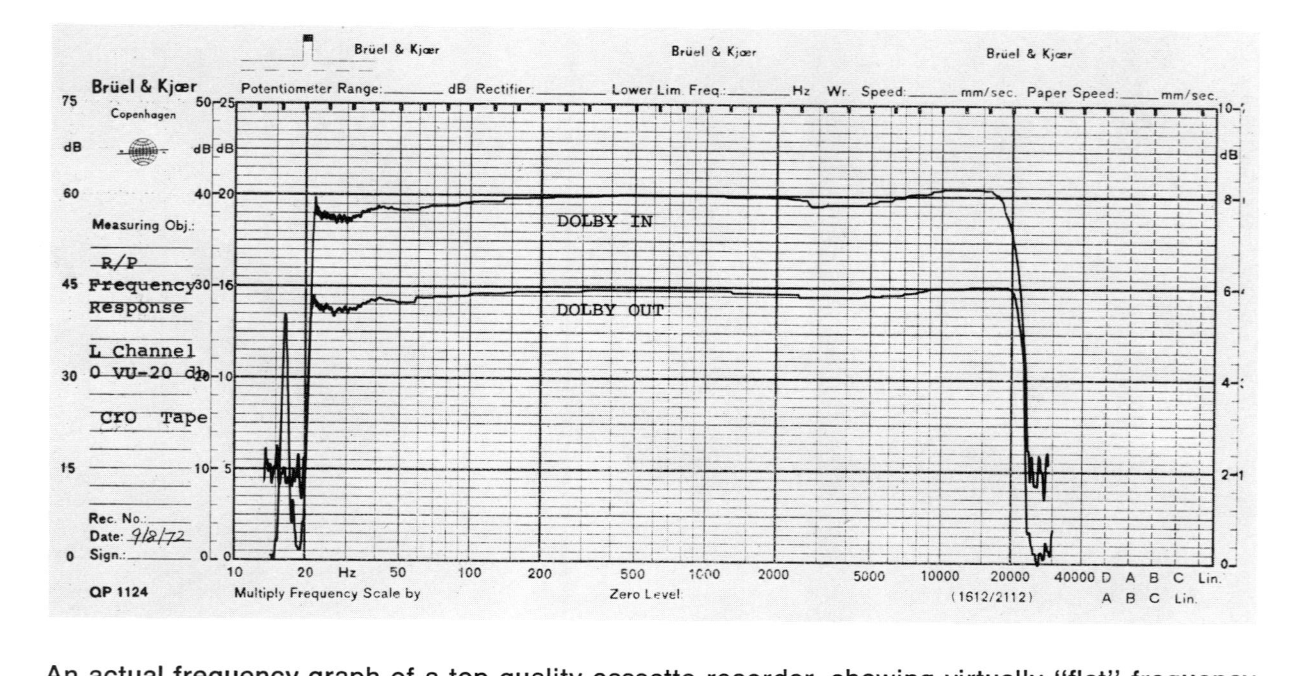

An actual frequency graph of a top-quality cassette recorder, showing virtually "flat" frequency response over the entire musical spectrum from 20 to 20,000 Hz, both with and without Dolby circuit in operation. (*Courtesy Concord-Nakam*)

meaning that no sound is more than 3 db louder (+) or softer (−) than it would be under ideal "flat" conditions. The term "flat" stems from the fact that if an ideal frequency response is plotted on graph paper, it comes out as a straight horizontal (i.e., flat) line. The plus or minus deviations, if shown as a graph, show up as humps or dips in the line.

The overall frequency range covered varies with tape speed, for reasons explained in the preceding chapter. Typical overall response figures for a good recorder might read as follows: 30 to 18,000 Hz at 7½ ips, 30 to 14,000 Hz at 3¾ ips, and 50 to 7,000 Hz at 1⅞ ips.

Distortion

Distortion usually varies with the intensity of the recorded signal. The amount of distortion stated in the specs is the percentage of distortion relative to the total sound output for a test tone recorded at the intensity level where the recording meter reads zero. (The zero point on the meter marks the highest sound intensity that can be recorded without greatly increased distortion.) Above zero level, the pointer of the recording meter swings into the "forbidden" zone—usually colored red on the meter scale—warning that distortion at those levels is likely to be troublesome. On a good recorder, total distortion should not exceed 2–3 percent at zero level, and the specifications should clearly say so. Of course, the kind of tape used also affects the amount of distortion produced at high recording levels, and the measurements stated in the specifications are generally made with tape of very high quality.

Flutter and Wow

These important performance factors are also expressed as a percentage. On very good open-reel recorders, flutter and wow should not exceed 0.1 percent, which means that the speed variations in the flow of the tape are so small as to be entirely imperceptible. The very best cassette recorders now attain comparable levels of performance, but on most of them, the figures for flutter and wow are somewhat higher than on open-reel machines. However, cassette machines with flutter and wow rated at 0.2 percent may still be regarded as very good indeed. Most of the inexpensive portable cassette recorders don't come anywhere near such standards, which accounts for their strangely tremulous, unsteady sound.

Signal-to-Noise Ratio

This important specification expresses the loudness difference between a recorded test tone and the extraneous noise added by the recorder during the recording process. Unfortunately, no standard conditions for making this measurement have ever been formulated. True enough, the National Association of Broadcasters (NAB) has proposed such a standard, but not all manufacturers adhere to it when testing their own products. As a result, the printed specifications of some inferior

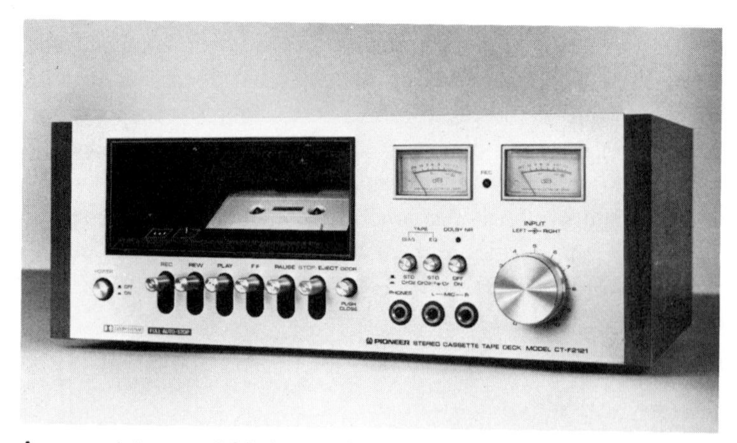

A recent type of high-quality cassette recorder featuring front-loading of the cassette, thus leaving the top free for stacking other audio components. (*Courtesy Pioneer Co.*)

recorders look a lot better—on paper—than those of their superior competitors rated by the more stringent NAB method. But if you see the signal-to-noise specifications followed by the letters NAB, you can be sure that the measurement is reliable and the manufacturer is describing his product honestly.

The NAB measurement of signal-to-noise ratio (usually abbreviated S/N) on the better machines is –50 db or more, which means that the signal you want to hear is 50 db louder than the background noise introduced by the machine. For all practical purposes this means that the background noise produced by the recorder on the tape (as distinct from inherent tape hiss) is virtually inaudible.

Special Features

In addition to making sure that these basic specifications measure up to the performance standards required for satisfactory results, you should also make sure that the recorder

This top-quality cassette recorder also features front-loading and also has built-in input mixing facilities as well as solenoid-operated controls. In performance, it equals open-reel standards. (*Courtesy Concord-Nakam*)

you buy has all the special operating features and conveniences you want. Among these, you might consider the following:

Automatic Reverse. This means that you don't have to flip the tape over at the end of a reel (or at the end of a cassette) to play the second pair of stereo tracks in the reverse direction. Instead, the machine automatically puts the tape in reverse so that the second side is played after the first. On most auto-reverse machines, this feature operates only in playback—not in recording. However, on a few open-reel recorders it works both for playback and recording. This has the advantage that if you reach the end of a reel while the program you are recording still goes on, you don't lose continuity while reversing reels. Even so, two or three seconds are lost during the automatic switch-over. In speech recording, this may not matter a great deal since only a few words are lost. In music, by contrast, the effect can be rather disheartening. I remember recording an opera broadcast off the air when my machine decided to do its automated flip-trick right in the middle of the tenor's magnificent high C. Which brings us to the next consideration.

Reel Capacity. Suppose you're recording a unique event—either "live" or broadcast. You know your recording has to be right the first time, because you'll never get a chance to do it over again. The last thing you want to happen is to run out of tape in the middle of it. You have to make sure you have enough tape loaded into your recorder to last the whole program—or at least to some intermission point which you can use to put on a new reel. In short, what you are concerned with is the uninterrupted recording time obtainable on your machine.

On cassette recorders, the longest uninterrupted recording span is one hour (which is exactly one side of a 120-minute cassette). Cassettes simply cannot contain a greater quantity of tape because their outer dimensions are fixed. True, some experimental cassettes have been produced with extra-thin tape enabling more tape to be wound up on the small internal reels of the cassette. But these cassettes have proved very unreliable because the extra-thin tape breaks or stretches too easily. For all practical purposes, 120 minutes per cassette remains the limit—which means 60 minutes of uninterrupted recording.

With open-reel recorders, by contrast, there is a choice of reel capacity. Virtually all open-reel recorders take standard seven-inch reels which hold up to 2,400 feet of tape. This provides an uninterrupted recording span of one hour at 7½ ips or two hours at 3¾ ips, which should be sufficient for most programs. Yet some of the fancier recorders take the large, professional-type 10½-inch reels, which contain up to twice as much tape and thus double the maximum uninterrupted recording time. Or else, the larger reel allows you to use a somewhat thicker type of tape and still have enough for the required time span, and the thicker tape usually gives you somewhat better dynamic range and greater break-resistance.

Monitor Facilities. We have already mentioned this feature in Chapter 4, pointing out that three-head design permits you to check your recording while you are making it. Such monitor facilities are standard on all the better open-reel recorders, but only a few of the very best cassette machines offer this feature.

Special Effects. Many open-reel recorders offer a feature called "sound-on-sound," which is based on the ability of the machine to record on one channel while playing the other. That way you can transfer a recording already made on one channel to the other channel and at the same time add new musical parts to the previously recorded performance. That way you can sing duets with yourself, add your own voice to a broadcast or to the music from a phonograph record, create echo

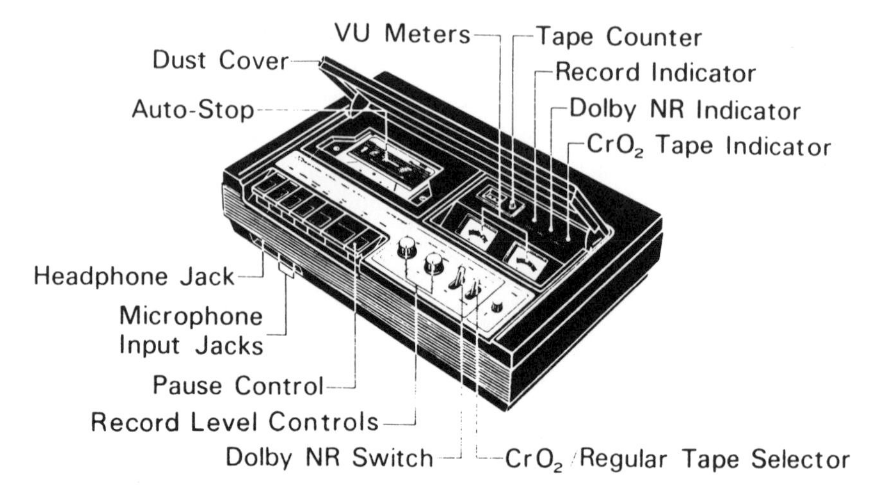

Features and controls of a typical cassette deck.

effects, add narration to background music or vice versa, or do any number of similar tricks. Such sound-on-sound facilities usually are offered only on open-reel machines, not on cassette recorders.

Some of the more elaborate cassette machines, as well as many open-reel recorders, feature built-in mixing facilities. This means that you have separate volume controls for two (and sometimes more) separate inputs. This allows you to set up separate mikes for a singer and for the accompanist and adjust the balance between them for optimum effect. Or you can use the separate controls to adjust the balance between different instrumental groups in a band or orchestra or for the separately miked members of a rock group. Such built-in mixing facilities are usually found only in the more expensive models. Yet even relatively low-priced units sometimes feature separate controls for the "line input" (i.e., the input from radio or phonograph) and the microphone input. This, too, provides added versatility. For example, if photography is another of your hobbies, the separate line and mike controls are handy in making sound tracks for your slide shows, blending your

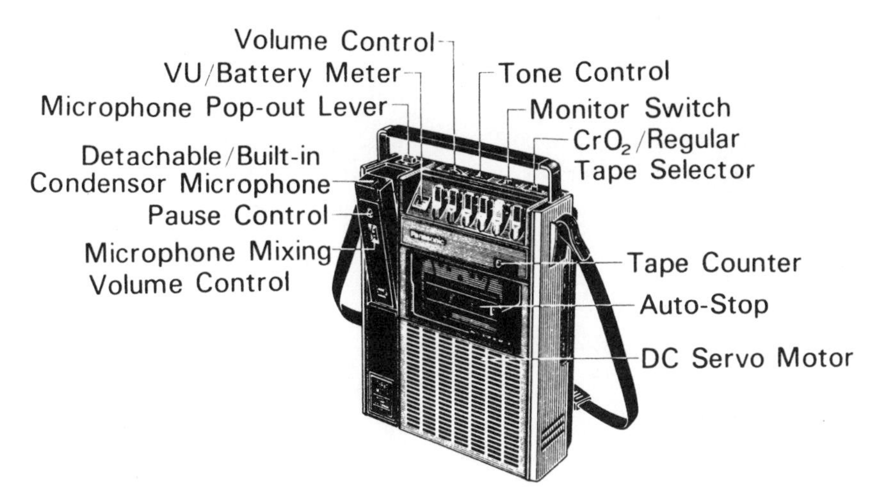

Features and controls of a high-quality portable cassette recorder. (*Courtesy Panasonic*)

"on-mike" narration with background music from a turntable.

By way of gilding the lily, some top-notch open-reel recorders also have a Dolby noise suppressor. As we have pointed out, the Dolby is usually found only in cassette machines, where it serves to reduce tape hiss and widen the dynamic range. Since the hiss level on open-reel tape is very low to begin with, the Dolby is usually considered unnecessary for open-reel recorders, but there is one particular application where is may be useful even in open-reel recording. If you use the sound-on-sound feature to add one track to another in multiple "overdubbing," the background noise builds up slightly with each dub. After all, you are transferring the tape hiss of the original recording to the new recording. By suppressing the tape hiss, the Dolby lets you do such overdubbing without increasing the background noise.

Still, if your recorder does not have a Dolby, and you find later that such a feature would be useful, you can buy a Dolby device as an accessory.

As a rule, elaborate special features on a recorder tend to

be costly. Unless you plan to do fairly sophisticated multi-track recording work, there is no point in paying for them. Many manufacturers offer decks designed as a sort of optimum compromise between cost and elaboration, geared to the needs of the typical user. In such models you will find good basic specifications, but few frills and special features.

One very convenient feature found on some cassette decks is known as a "memory." Its purpose is to locate automatically a particular place on the tape. As the machine plays, it may come to a selection you may want to hear again. In that case, all you have to do is push a button. This activates the "memory," and when you rewind the cassette later, it will stop at exactly the point you marked to replay the section you wanted. But even without the memory feature, you can locate different parts of the tape by watching the numbers on the tape counter.

In conclusion, a few general observations should be made about the numerical specifications listed earlier. If you compare the specifications of different makes and models, you may find that outstanding performance in one respect is sometimes obtained by compromising some other performance factor. This kind of trade-off of one quality for another is encountered in many different fields of engineering. For example, in automobile design, a car may have a very soft ride attained at the cost of poor handling, or it may have outstanding fuel economy at the cost of low power and poor acceleration.

The same kind of engineering compromises sometimes enters into the design of tape recorders. For example, signal-to-noise ratio or distortion may be compromised in the interest of wider frequency response. As a practical guideline, this means that a tape recorder with conservative specifications is probably a more honestly engineered product—and therefore a better overall performer—than a competitively priced machine with suspiciously superior claims.

Careful quality control in manufacturing is the key to satisfactory performance. Here a Teac cassette deck is checked out in laboratory.

Lastly, keep in mind that the specs are, at best, only a partial description of a tape recorder's total performance. A tape machine is, after all, a curious mixture of three separate realms of technology—mechanics, magnetics, and electronics. The interaction of these three factors involves so many variables that the specs alone can't tell the whole story. They merely serve as guideposts. The only true test of a given model is operational—the direct impression made by its sound and its "feel."

A well-known audio designer once quipped that the specs can no more define the total performance of a tape recorder than a recipe can convey the exact flavor of a dish. Still, from reading the recipe you can get a pretty good idea of what the dish might be like. The specs have a similar degree of utility. Though they may be neither complete nor unequivocal, they still provide a fairly objective framework for evaluating the product. They help you determine which recorders meet essential quality factors and offer the operating features you want. From this group you can then make your final selection by actually putting the machines through their paces, finding out how the controls "feel" as they respond to your touch, and assessing the sound with your own ears. The proof of this particular pudding, to revert to our culinary metaphor, lies in the listening.

CHAPTER 6

MICROPHONES —THE EARS OF YOUR RECORDER

For making "live" recordings, the right kind of microphone is the key to success. After all, the mikes are the ears of your machine. What they "hear," you'll hear. Since they are the gateway by which the sound enters into your recorder, they occupy a crucial position in the whole recording process. If the microphones limit the quality of sound before it even reaches the recorder, nothing can make up for the loss later.

Cheap portable recorders usually come with mikes supplied by the manufacturer. Most such mikes make no claim to quality. They do well enough for casual recording of speech, but they often lack the range to catch the finer aspects of timbre and articulation of the human voice, and they are usually incapable of doing justice to music. Some portable recorders come equipped with built-in microphones. They are useful for recording mike-shy people simply because they don't look like microphones, being set into the housing of the recorder. Some of these built-in microphones are fairly adequate, but, being part of the recorder itself, they often pick up vibrations from the drive motor, which show up on the recording as a rumbling background noise. For this reason, a separate microphone is usually preferable, even for small portables.

The better tape decks, by contrast, seldom come supplied with their own mikes. Here the choice of the mike is left to the user, and often this choice becomes a highly personal matter, for the tonal character of microphones is quite individualistic.

Like musical instruments, different kinds of microphones tend to have their own quirks of tone coloration, so the choice between them becomes a rather subjective and sometimes even emotional affair. I have known recording engineers who swear by their particular pet mikes and speak of them as a violinist might speak of a prize fiddle.

The comparison is by no means arbitrary. In fact, microphones differ from each other in much the same way as one violin differs from another, and for similar reasons. Both in a fiddle and in a mike, the tonal character depends on the inherent resonance of the vibrating parts, and they are difficult to predict or control. Yet thanks to modern manufacturing techniques, these factors can now be controlled more effectively in manufacture, and current models of high-quality microphones can be trusted to perform reliably.

In essence, the microphone acts as a sort of middleman between two separate realms of nature: the realm of sound and the realm of electricity. It must translate the mechanical vibrations of sound into their electrical equivalents. As in other kinds of translation, something is often lost or altered in the process. To make the translation as accurate as possible is the task facing the microphone designer. Until recently, the complexity of this task made really good microphones quite expensive, many of them costing several hundred dollars apiece. Yet within the last few years, some radical improvements in microphone construction now make it possible to obtain mikes of remarkable fidelity in the price range between $50 and $150.

Basic Microphone Types

Among the half-dozen different kinds of microphones, only two basic types are now in widespread use for recording purposes. One is the so-called dynamic microphone, the other is the condenser microphone or capacitor microphone, as it is sometimes called.

The dynamic microphone works like a loudspeaker in reverse. Instead of converting electricity into sound, it converts sound into electricity by exactly the same means. A loosely suspended diaphragm picks up sound vibrations from the air. A coil of very fine wire attached to one part of the diaphragm moves within the field of a magnet. The moving coil in a magnetic fields acts like a miniature dynamo and generates an electric current proportional to the movement of the diaphragm, and hence proportional to the sound waves in the air moving the diaphragm. The frequency of the current so generated represents the pitch of the sound, and the strength of the current represents loudness. In this way, the dynamic microphone translates sound into an electric signal, which is fed to the tape recorder. Dynamic microphones of this type have the advantage of being relatively rugged, trouble-free, and inexpensive—all of which accounts for their popularity.

The other type of microphone—the condenser microphone—operates on a different scheme. Unlike the dynamic microphone, it does not generate an electric current. Instead, it modifies a current running through it in accordance with the motion of a delicate metal foil which swings with the sound waves of the surrounding air. Since very little mechanical force is required, the metal foil can be made very light and its motion can remain minimal. As a result, the condenser mike responds easily to elusive high frequencies, thus catching the subtlest overtones of musical instruments and lending crispness to the articulation of human speech. Because of this, the condenser microphone has become the favorite among professional recordists. Yet it is not without some drawbacks.

For one thing, condenser microphones are more complex and delicate than dynamic mikes. They require a built-in electronic circuit to supply their operating current and to boost their signals right at the source. This also makes them more expensive than the simpler dynamic microphones, though several good condenser microphones can be had for about $100.

A good microphone is the key to satisfactory "live" recordings. The group of microphones shown here includes both condenser and dynamic types, both omnidirectional and cardioid. (*Courtesy AKG*)

Oddly enough, the very virtues of the condenser microphone may create problems for inexpert recordists. Because of their enthusiastic response to those sparkling highs and subtle overtones, condenser microphones have a tendency to make the sound of some musical instruments over-brilliant. For brass and woodwinds, this doesn't matter much. In fact, it lends edge and bite to the sound which many wind players like—and drummers invariably love. But when it comes to recording string instruments, it's quite another matter; for the high-frequency characteristics of many condenser mikes make string instruments, especially violins, sound hard and glassy, and even an accomplished fiddler comes out scratchy. For the same reason, the human voice sometimes seems overly sibilant.

These difficulties are easily neutralized by placing the mikes at just the right angle and distance from the sound source, and by doing just that, experienced recordists can get superb results from condenser microphones. The chapter on "Tips and Techniques for Live Recording" will give some hints on

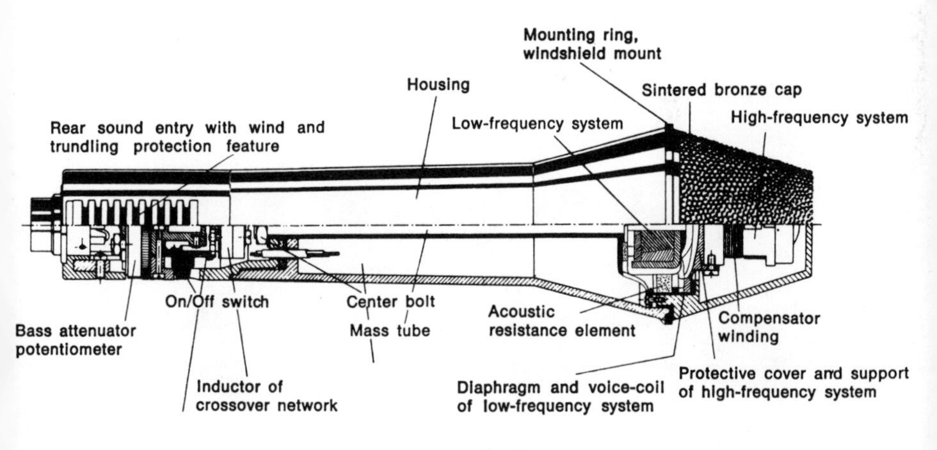

Cutaway drawing shows the construction of an AKG microphone with separate elements for high and low frequencies.

recommended patterns of mike placement, but on the whole, such things are best learned by trial and—alas—error. For beginners, therefore, a good dynamic microphone is easier to use, being more forgiving of mistakes in placement and likely to give more consistently pleasing results.

How Mikes Hear

Human ears hear in all directions. You are aware of sounds coming from the rear and from the sides as well as from the front. In short, human hearing is omnidirectional. Most microphones are built with the same omnidirectional characteristic. They pick up sound from all over.

Yet there is another aspect to human hearing which is mainly psychological. It's a kind of mental switch that lets you tune out what you don't want to hear and concentrate on what's important. At a concert, for example, you can ignore coughs and sneezes as you focus on the music. Or at a noisy party, you can mentally blank out the surrounding hubbub to concentrate on the person who, for whatever reasons, has engaged your attention. In fact, psychologists studying this phenomenon have called this capacity for selective hearing the "cocktail-party effect."

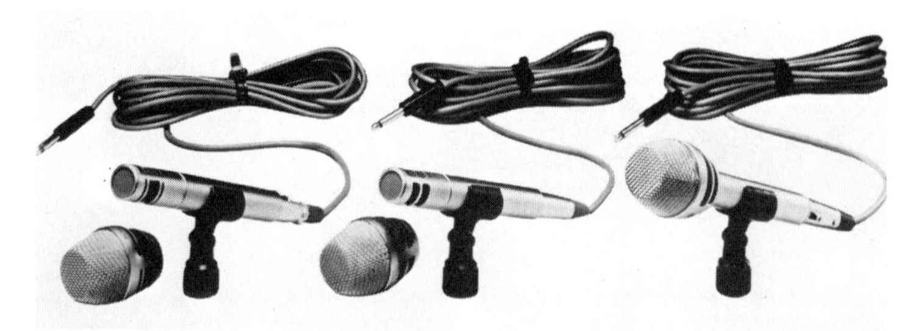

Many microphones feature detachable wind-screens to shut out wind noises in outdoor recording or to attenuate the breathing sounds on such consonants as *t* and *p* when used close up to a speaker or singer. (*Courtesy Technics*)

To simulate this aspect of human hearing, engineers have designed so-called directional microphones. These microphones, so to speak, concentrate their attention in a chosen direction. The most common type "listens" for sounds from the front, ignoring sounds from the rear. Such mikes are useful in recording live performances in noisy locations—at a nightclub or in some similar setting. In that case, the mike "listens" to the singers or the musicians while shutting out audience noises. Mikes of this kind are also handy for minimizing echo in an overly reverberant room, for they reduce the sound reflected from the wall. This often helps to increase the clarity of speech, particularly when recording in a large hall or in a church.

Polar Patterns

The directional characteristics of a microphone can be clearly shown by its polar pattern. This is a graph showing

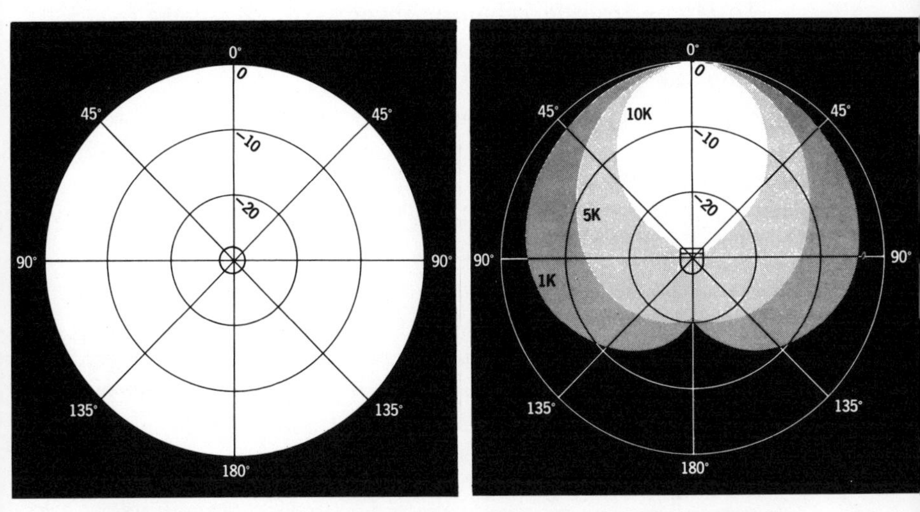

(*Left*) Microphones come with different pickup patterns. Omnidirectional microphones accept sound from all directions. (*Right*) Cardioid microphones favor sound from the front, blocking sound from the rear. The two graphs show their respective patterns. (*Courtesy AKG*)

the microphone's sensitivity to sounds arriving from various directions. According to these patterns of sound acceptance, microphones are classified as omnidirectional (accepting sound from all around), cardioid (accepting sound mostly from the front), and bidirectional (accepting sound from front and rear while suppressing sound from the sides). Among directional microphones, the cardioid is the most widely used. The term cardioid literally means heart-shaped, because the polar pattern of this microphone resembles the drawing of a heart, with the tip representing the favored direction of sound acceptance. The bidirectional mike with its front-and-back pattern has only limited usefulness in general recording. It is mostly used in TV studios for broadcasts of two persons talking to each other across a table. In that case, the pickup pattern of the mike covers the two speakers while shutting out surrounding sounds.

Typical Uses

To form a realistic judgment about the kind of mike most suited to your particular recording projects, let's consider a few typical situations.

Suppose you are recording a band or orchestra in stereo. In that case, a pair of cardioid mikes aimed at different sections of the band will give you nice stereo separation and, at the same time, hold down unwanted audience noise. What's more, you can adjust the amount of reverberation by moving the mikes closer or farther away from the sound source. The farther away you go, the more reverb you'll catch.

Suppose, on the other hand, you're recording a folk singer with his guitar. In that case, you might place a couple of omidirectional mikes very close, so that one picks up mainly the voice while the other concentrates on the guitar. This gives a nice effect in stereo playback while the omnidirectional pickup prevents complete channel separation between voice and guitar—which would seem quite unnatural. Besides, the omnidirectional mikes will also pick up enough ambient reverberation to make both the singer and the guitar sound resonant and full-bodied, despite the close miking.

I have found a cardioid microphone very helpful in a classroom recording situation where I was not allowed to set up a microphone up front on the lecturer's desk. Still, I got good results with a portable cassette recorder, sitting several rows back in the class and holding the cardioid mike in my lap. Thanks to its polar pattern, the mike was "reaching forward" toward the speaker, pulling him acoustically close, while shutting out much of the reverberation from the plaster walls, that would otherwise have blurred his speech.

Mikes on a Budget

The ideal situation, obviously, is to own two pairs of microphones, one omnidirectional and one cardioid. That way you could always choose the mikes best suited to a particular recording task. But what if you are on budget and can afford only one pair of mikes? What you are looking for, in that case, is a pair of economical mikes with maximum versatility.

To help in making this selection, let's first consider what *not* to get—and why not. First, let me suggest that you take a skeptical attitude toward condenser microphones. Not only are they more expensive, but for reasons already explained, they must be used with experienced discretion. All things considered, a pair of good dynamic mikes, selling between $50 and $100 apiece, would be your best bet for starters.

Next you have to decide whether to get omnidirectional or cardioid models. Personally, I like working with cardioids because they allow me to pinpoint the sound source a little more clearly in a variety of acoustic surroundings. But a pair of omnidirectional mikes may well prove more versatile and suitable for general use. The omnidirectional mikes provide more of the sense of the acoustic spaces of the original performance, and this sense of ambience is often just what is wanted. Moreover, the ambience factor can be easily varied by adjusting the distance of the mike from the sound source. Cardioid mikes, by contrast, tend to isolate the performers from their acoustic surroundings, sometimes resulting in a rather "dry" sound.

The omnidirectionals—or "omnis," as they are called for short—have another important advantage. Unlike the cardioids, you can move them very close to a performer without running into a problem known as the proximity effect. This effect, inherent in the construction of cardioids, causes them to change their frequency response at very close range, making the sound very boomy. If you like miking a musician very close—the preferred technique for a lot of solo work—you'll have an easier time with omnis.

All this adds up to a specific recommendation: If you don't have much prior experience with mike placement and if you want to start out with just one pair of mikes, the most suitable, versatile, and economical mikes you can get is a pair of omnidirectional dynamics. There is a wide choice of makes and models, and the products of such companies as AKG, Shure, Electro-Voice, Advent, Beyer, Sennheiser, and Sony can be highly recommended.

Mike Specs

Picking a mike, you'd naturally check out the specifications, the most important of which is frequency response. No microphone is entirely "flat" in its response. In fact, the frequency graph of even a good microphone may look like a profile of the Rocky Mountains, and it is precisely these erratic response deviations that give the mike its individual character. Such response graphs are furnished with only the better mikes. But in any case, you should look for the "plus-minus figure" in the specifications which states the maximum deviation from the ideal flat response. If the deviation is no greater than ±4 db, you can take it for granted that the mike sounds pretty good. As for overall range, very inexpensive mikes may have a response from only 100 to 10,000 Hz, but if you plan to make only speech recordings, this is sufficient. For musical recording, the response should be at least 50 to 14,000 Hz in a medium-priced mike, or for top quality, at least 30 to 15,000 Hz.

The finest professional microphones have wider frequency response at both the high and the low end, and some of them even feature a switch to let you alter their directional characteristics, choosing either omnidirectional or cardioid patterns. But such microphones are very expensive, costing several hundred dollars. Except for professional studio applications, this expense seems hardly justified in view of the very good results obtainable with microphones selling for between $50 and $100.

Impedance and Sensitivity

Aside from frequency response, microphones have two other main specifications. One of these is impedance. The term describes the degree to which a circuit impedes the flow of alternating current. If two devices in an audio system are to be connected to each other, their impedance must be similar. Otherwise, the flow of electrical energy is blocked. Impedance is often abbreviated by the symbol "Z" and expressed in ohms (the unit of electrical resistance). Microphones are rated either as low-impedance (low-Z) if their impedance is less than 600 ohms, or high-impedance (high-Z) if their impedance is greater. The impedance rating has nothing whatever to do with the quality of the mike as such. It merely tells you whether the mike can be matched to your recorder. Virtually all modern tape recorders are designed for low-impedance mikes, and most modern microphones are of this type. If your recorder is less than five years old, chances are that it requires a low-impedance mike. But it's a good idea to check the specs, and sometimes the recorder manufacturer recommends certain types of microphones that best match the machine.

If you happen to have a high-impedance mike and a low-impedance recorder (or vice versa) there are still ways to make them compatible. All you need is a microphone transformer (available separately) to plug into the connecting line between mike and recorder. The transformer converts high-impedance to low-impedance (or the other way round) and thus can make any mike compatible with any recorder.

If you are using a high-impedance mike, you should be aware of an inherent drawback: if the connecting cable between mike and recorder is longer than 15 or 20 feet, some of the high frequencies get lost along the way. The longer the cable, the greater the loss. That's one of the reasons why modern recorders are set for low-impedance microphones, whose cables can be much longer without causing any frequency losses. This is very convenient in "live" recording situations if you want to set up your mikes near the performers but want to keep the recorder out of sight some distance away.

But even the preferred low-impedance mikes may occasionally give you cable problems. If the cable runs near other electric wires carrying standard AC current, the mike cable might pick up some hum from them. Usually you can remedy this difficulty just by shifting the microphone cable a few feet to another location, further removed from the source of the electrical disurbance.

The sensitivity rating of the microphone tells you how much signal the mike will produce for a certain amount of sound pressure acting upon it from the surrounding air. A typical specification might read "–60 db re 1 volt/ 1 microbar"—all of which is highly technical, confusing, and not really very informative because there is no standard way of measuring sensitivity. It suffices to remember that the smaller the decibel number, the more sensitive the mike. For example, a mike rated at −70 db is less sensitive than one rated at −60 db.

For all practical purposes, you can ignore the sensitivity rating. Virtually all mikes provide sufficient output to drive your tape recorder, and whatever difference exists between various models you can easily adjust for with the gain control of your recorder. If you happen to have a slightly less sensitive mike, all you do is turn up the recording volume a little.

Caring for Your Mike

Microphones require tender loving care for survival. Some are sturdier than others, and we have already mentioned that dynamic mikes, on the whole, are more rugged than condenser mikes. But by their very nature, all microphones—except the knockabout low-fi types used by cops and taxi-drivers—are delicate instruments. After all, their moving parts must respond to the most subtle vibrations. This necessarily entails a certain fragility of their mechanical structure. Compared to the moving parts of wide-range mike, the wings of a butterfly would seem massive. As a result, a sharp knock can disable a microphone, and a fall to the floor is usually fatal. It is a good idea to handle microphones as if they were made of glass, and they should be transported only in the special protective cases furnished by the manufacturers. Above all, avoid the common but atrocious practice of tapping the mike with your fingers or blowing into it to see if the mike is switched on. Such rude gestures seem like earthquakes and hurricanes to the inner workings of your mike, which are attuned to the faintest tremors of the air. If you want to check whether any sound is coming through your mike, just speak to it gently.

CHAPTER 7

THE RADIO BONANZA —TAPING OFF THE AIR

With the advent of radio, music has become a kind of natural resource—one of man's less lethal additions to the atmosphere. You can tap the great musical reservoir in the sky and lay in your own permanent supply of music simply by taping it off the air. What's more, it's free. Almost, that is. You still have to pay for the tape. But compared to phonograph records, it's a bargain.

An hour's worth of music on records—calculated by the average 40-minute playing time of a standard-price LP record—runs almost $9.00. If you tape your own stereo cassettes off the air, you cut the cost by more than half even if you use premium-quality cassettes. With economy cassettes, you might lose a pinch of those uppermost highs, but the cost-per-hour comes down to about $2.00. And with open-reel tape, running at 3¾ ips, it works out cheaper yet.

But money isn't everything. Consider variety. Whatever your taste—classics, jazz, folk, rock, or pop—with more than a thousand FM stereo stations dotting the U.S. map, chances are you can find what you like on the air. Especially if you live in or near any of the larger cities, where FM stations are clustered, all kinds of musical pleasure floats over your roof.

Nor need you limit yourself to music. Particularly if you are within range of the Public Broadcasting System, you may find other kinds of programs worth preserving for your personal tape library. You might start a collection of radio drama—plays ranging from Shakespeare to Ionesco. Or you might want to keep a "live" file of talk-shows with famous personalities, lectures of enduring interest, sports events for your own Hall of Fame. You might start your own archive of Presidential pronouncements, and by taping important newscasts and on-the-scene reports, you can put together your own version of "recorded" history.

In fact, you can take a chance at taping any program that promises to be interesting enough to hear again; for—unlike a phonograph record—a tape recording is not necessarily a permanent investment. If you find the program disappointing, just use the same tape next time to record something else. I have built my own collection on such a trial basis, discarding and reusing as much as 80 percent of my "takes." For example, I recorded all kinds of music, from symphonies to acid rock—just to get better acquainted with the individual pieces through repeated hearing. If I grew tired of a number, I just used the same tape for another musical exploration. That way, my tape deck became for me a prime instrument of musical discovery, and thanks to this flexible approach, my tape library is changing to keep pace with my often fickle taste.

In some ways, the musical choice on the air is even wider than on records. Many classical FM stations, for example, carry live concerts with soloists, orchestras, and conductors that could never appear together on records simply because they are under exclusive contracts to different record companies. The Public Broadcasting station in my own area, the Berkshire Hills of Massachusetts, features broadcasts from the world's major music festivals. Because of the commercial structures just mentioned, most of these unique performances could never find their way onto disks. But they have become part of my personal tape collection.

Killing and Keeping Commercials

In recording music off the air, I usually cut out the commercials. When I used an open-reel recorder, I did so quite literally, snipping out the parts of the tape which carried the commercial plugs. Lately, I have been doing most of my off-the-air taping with a cassette deck, which does not permit subsequent editing. So I keep a finger ready on the PAUSE button when a piece of music nears its end, halting the recording before the announcer's voice comes on. By stopping and starting the tape instantly—while leaving all other control settings intact—the pause control lets you edit your tapes electronically as you make them, deleting unwanted portions of the program.

A friend of mine takes a different approach. He actually collects commercials. He keeps his cassette deck loaded and ready to go, and turns it on whenever his favorite stations get ready to run off a string of commercials, usually just before and after the hourly station-break. Today's better commercials, he says, are gems of creativity with an amusement value often greater than that of the programs they interrupt. Frankly, I think he overstates his case. I am not willing to grant commercials the status of an art form, but I admit that some of them are musically delightful and performed with polish and zest, and perhaps they are significant as a sidelight on contemporary cultural history. At any rate, you may find that these cheerful reminders about Carlsberg Beer, Vita Herring, Volkswagen, and whatnot have a certain entertainment value when collected over the years, and you might want to make them part of your airborne bonanza.

The Basic Hookup

As audio editor of a national magazine, I was surprised to find out from my reader mail that some people still record radio programs simply by putting mikes in front of their loudspeakers. This obvious, if primitive, method hampers recording in three ways: 1. Your dog, dishwasher, telephone, spouse, and the car honking out in the street will all be on your tape along with the music. Family mementos are fine, but not as background for Beethoven. 2. The acoustics of your living room sneak into the mikes and are superimposed on the original acoustics of the studio where the program originates. The result of this acoustic double vision is blurred sound on the tape. 3. The limitations of your mikes and loudspeakers become part of the recording.

The preferable alternative is to bypass your loudspeakers, your mikes, and your living room entirely and eliminate these drawbacks. You do this by hooking your recorder to your radio receiver by a direct electric link. Assuming you already own a good stereo system with an FM receiver or tuner, making these connections is very simple. All you need is a couple of connecting cables, obtainable at almost any radio shop.

At the rear of your receiver (or at the rear of your amplifier, if you're using a separate FM tuner) you will find two small round sockets marked TAPE OUT—L and R, the final two letters signifying Left and Right channels. Your connecting cables will have pin plugs at the end to fit into those small round terminals, and you should run two cables from the TAPE OUT terminals of your receiver to the input terminals of your tape deck marked LINE or PHONO. That's all. From then on, anything that comes in on your FM receiver is automatically piped to your tape deck, and you're set for recording.

For playback, use the same principle in reverse. From the output connections of your tape deck, run another pair of connecting cables to the terminals on your receiver or amplifier marked TAPE IN or MONITOR, again connecting left and right to the correspondingly marked terminals. Then, when you play your tapes and set the program selector switch on your receiver on TAPE or MONITOR, the sound will be reproduced through your stereo system. Once you have your deck hooked up in this way, it's a good idea to leave all the cables permanently connected. You can neatly stash them out of sight, and if the connection is kept, you're always ready to roll whenever an interesting program comes on the air.

Aside from assuring optimum tonal quality, this type of

hookup has another advantage: You can record without actually listening. In fact, as I shall explain later, you can even record without actually being there. All you have to do for silent recording is to turn down the volume on the receiver so that no sound emerges from your speaker. Yet if you do this, the signal still flows silently into your recorder to be replayed at a later time. Some of the most interesting radio programs are perversely scheduled just when I am too busy to listen. In that case, I let my recorder run silently, taking down the program while I go about my business. Later, at my leisure, when I can listen with full attention and enjoyment, I play the tape. In that way, I rearrange broadcast schedules to suit my personal convenience.

Only a high-quality stereo receiver should be used as a signal source. Ordinary FM radios, aside from lacking the proper output connections, rarely provide a high-quality, distortion-free signal—and you just can't get a good recording from a bad signal. If, in an emergency, you must use an ordinary radio or TV set as your sound source, you can plug your tape recorder into the earphone jack to tape the signal. However, be sure to turn down the volume control of the radio (or TV set) nearly all the way, so as not to overload your recorder. Under such conditions, the best way to adjust volume level is to turn the recording level control on your recorder to the midpoint. Then, starting from the lowest point, gradually turn up the volume control on the radio until your recording meter indicates the normal signal level with the meter hitting zero at the loudness peaks.

The reason why FM is the preferred signal source for off-the-air recording is that AM is inherently incapable of adequate fidelity. It is limited in frequency range, cutting off all the highs beyond 5,000 Hz, is beset by static, and has a restricted dynamic range. Essentially, AM is merely a survival from the early days of radio, and only FM broadcasts with the kind of fidelity that makes a recording satisfactory. Besides, FM is the only broadcast medium capable of stereo. Of course, when recording speech, newscasts, or other nonmusical material, neither fidelity nor stereo matter so much. For that type of broadcast, you may be satisfied with the quality attainable on monophonic AM or the audio portion of TV.

Getting Set

The secret of successful recording is a little advance preparation. Suppose the broadcast you want to catch goes on at 8:30. In that case, you had best be on the job about ten minutes ahead of time, just to make sure all systems are go. Record a minute or so of the preceding program, just to check out the control settings and make sure you're getting the broadcast on the tape. Then roll back the tape to the beginning and wait for the program you really want to record.

Be finicky about tuning in your station. It's got to be right on the nose—otherwise the sound will be fuzzy. Don't rely entirely on the tuning meter of your receiver to tell you when the station is tuned accurately. Always confirm the meter reading by rocking the tuning knob back and forth until you get the clearest sound.

Above all, make sure you have enough tape loaded on your deck. Nothing is more exasperating than running out of tape right in the middle of recording a broadcast. How long a tape will run depends, of course, upon two factors—the length of the tape and the speed of the recording. Following is a basic table for open-reel tape timing.

Recording One Direction

LENGTH	*1⅞ ips*	*3¾ ips*	*7½ ips*
150 ft.	15 min.	7½ min.	3¾ min.
225 ft.	24 min.	12 min.	6 min.
300 ft.	30 min.	15 min.	7½ min.
600 ft.	1 hr.	30 min.	15 min.

(*Continued on next page*)

Recording One Direction (*continued*)

LENGTH	*1 ⅞ ips*	*3 ¾ ips*	*7 ½ ips*
900 ft.	1½ hrs.	45 min.	22½ min.
1200 ft.	2 hrs.	1 hr.	30 min.
1800 ft.	3 hrs.	1½ hrs.	45 min.
2400 ft.	4 hrs.	2 hrs.	1 hr.
3600 ft.	6 hrs.	3 hrs.	1½ hrs.

The radio guides in your local newspaper may give you some notion of the time allotted to a particular program. If you are recording pop music or jazz, you won't have much of a problem with the tape supply, for the selections are short, and you can flip your cassette or reel between numbers when you get close to the end. With classical music, it's a good idea to anticipate the duration of the piece you are about to record. For standard repertory, the period of composition provides a good clue. Early symphonies—such as those by Haydn, Mozart, or Schubert—usually fit quite comfortably into a 30-minute span. Later symphonies, like those by Brahms or Tchaikovsky, usually take no more than 45 minutes, as do most of Beethoven's. But a symphony by Mahler or a single act of a Wagnerian opera may take twice as long. If you don't have enough tape to record the whole work on a single run, you can flip tapes or cassettes between movements of the symphony or scene shifts of the opera. If you must change to a new reel or cassette, you can usually do this in a matter of seconds with a little practice, so these changes can also be made within the natural intermissions of the music. But be sure to have that extra reel or cassette ready, right next to your recorder, for the crucial moment when you need it.

Next, adjust recording-level controls to the stations you'll be taping, using as your test signal whatever program happens to precede the one you want to record. Most stations keep their signal level fairly constant from one program to the next, so if the recording level is set correctly for one program, chances are it will also be correct for the program to follow. Only commercials are often aired at a louder level than the rest of the programs—an annoying, illegal, but fairly widespread practice. So don't adjust your controls while a commercial is on.

To make sure that your two stereo channels are properly balanced, switch your FM receiver temporarily to mono. That assures that exactly the same signal appears in both channels, and you can therefore adjust for identical meter readings for both left and right channels. Then throw the switch back to stereo—and you're ready to go.

Reading the Meter

The most important single factor in getting a good "take" is to adjust the recording level correctly. This means developing the right "feel" for reading the level meter. Most meters are calibrated in decibels on a scale that runs from about −20 db through zero to +3 db. The plus part of the scale is usually marked in red, and it's forbidden territory: if you want clean sound, the pointer should never go there. The usual working range of the meter for musical recording lies between −15 db and 0, but the actual numbers don't really mean much for practical purposes. What's important to remember is that zero is the upper limit. If the pointer swings beyond zero, too strong a signal is pushed on the tape and the sound becomes distorted due to what is called magnetic saturation. This simply means that the magnetic force becomes too great to be adequately absorbed by the metal particles on the tape. The magnetic flux then spills out of control and the sound gets all hashed up. So keep your recording level below zero, even though the average loudness of the music may not register more than about −7 db on the meter. These hints, by the way, apply not only to off-the-air taping but to all kinds of recording.

Reading the recording meter.

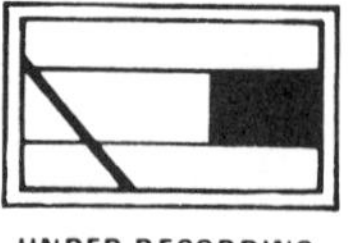

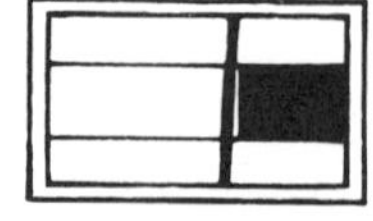

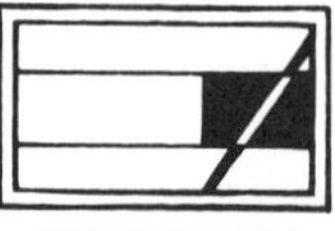

CORRECT RECORDING LEVEL

OVER-RECORDING

a. On many small portable machines the meter is not calibrated in numbers but color-coded. To avoid distortion, the indicator should not enter the black portion of the scale.

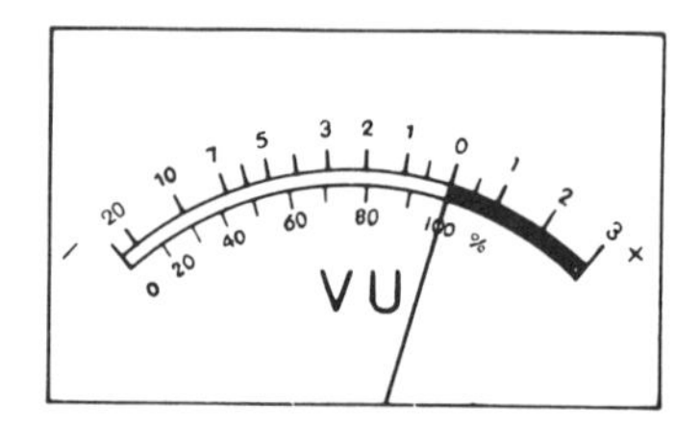

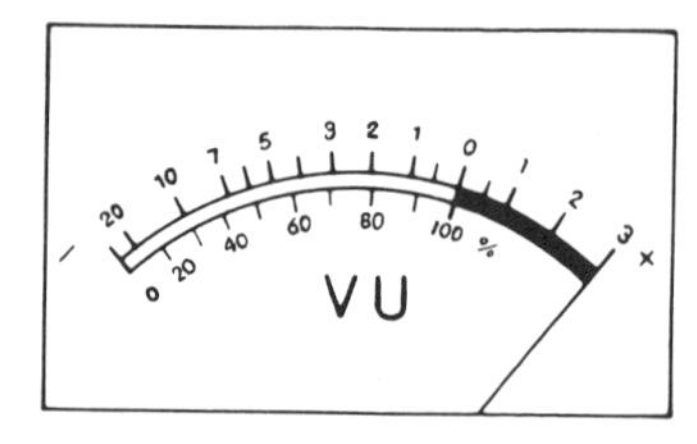

b. On most standard decks, the meter is calibrated in db or "volume units" (VU). The level should not exceed the zero point, as shown at left. Excessive level, as shown at right, will produce distortion. (*Courtesy 3M Company*)

Suppose you have your recording level set carefully before the start of your program by taking a meter reading on the preceding broadcast. Normally, everything should work fine from there on, providing the engineer at the station is doing his job properly. But if the engineer is asleep at the switch, you may find after your program starts that the level is either too high or too low. What then?

Naturally, you must then readjust the level while the recording is in progress. This is known as "gain riding" and has to be done quite discreetly. Nothing is so jarring as a sudden volume change, and an abrupt adjustment could ruin your recording. So turn the knobs very slowly, very slightly, and turn both channels simultaneously. A good trick in gain riding practiced by veteran engineers is to follow the shape of the music. This simply means that, if you must raise the level, do it during a crescendo, hiding your maneuver under the natural swell of the music. Conversely, lower the recording level during a passage where the music naturally subsides. But keep such corrections at a minimum. Frequent volume changes, even when skillfully accomplished, destroy the expressive content of the music because they interfere with the performers' own phrasing. The best recordings are those made with the least amount of knob-twiddling.

A final suggestion for recording "live" musical broadcasts: don't snap off the recorder at the last note. You might amputate the fading echoes of the last chord, ending an otherwise fine performance with a jolt. Wait for the applause, then slowly fade out both channels in tandem. That way you retain something of the atmosphere of the concert.

It may take you a while to get the feel of your level meter, for the meters on different recorders react in their own particular ways to the heavings of the music. Some meters are engineered to respond almost instantly to every beat of the music. This provides the most accurate indication of signal strength at any given moment, but it makes the pointers jump around so much that the meter is hard to read. That's why most machines designed for home recording use a delay circuit to smooth out the motion of the pointer. In effect, such a meter averages volume readings over a brief time span. In addition, some of the fancier tape decks feature a separate peak-indicator light that flashes a warning when the level gets too high.

If you like to let your recorder run unattended while doing something else, you might find a machine with an automatic peak-limiter very helpful. The limiter automatically reduces the signal level when the music gets too loud. Of course, this also reduces the dynamic range and may rob the loudness peaks of some of their impact, but if the basic recording level is properly set, these automated peak cutbacks will probably turn out to be so slight as to be hardly noticeable. In any case, such limiting is less objectionable than the ear-grating distortion that sometimes besets unattended recordings at the moments of musical climax.

So far, I have inveighed against the perils of recording at too high a level. What about recording at too low a level? The only danger there is that the background hiss becomes more noticeable during the soft passages, which is far less distressing than peak distortion. It is therefore preferable to allow a margin of error on the low side rather than the high.

Absentee Recording

If you have to be away from home when a program you want to record comes on the air, you still don't have to miss it. You can plug the power cord of your receiver into a clock timer and then connect the power cord of your tape deck to the switched power outlet at the rear of the receiver. Or you can connect both the receiver and the recorder to the clock timer. Set the timer slightly ahead of broadcast time, and both the receiver and the recorder will then be running to catch your program. Of course you must tune the station in advance, set your recording level, and leave the power switches of both receiver and recorder in the ON position. Also, the controls of the recorder must be locked into the RECORD position. Most recorders will shut themselves off automatically at the end of a reel or a cassette. As for the receiver, it doesn't matter if it stays on until you get home—and then the "missed" program will be waiting for you.

Dolby Broadcasts

We have already mentioned the Dolby as a device to reduce background noise in cassette recorders. Lately, the Dolby has also been employed by FM stations in their transmitters to reduce background noise in fringe reception areas.

If the station you are taping is broadcasting a Dolby signal (find out simply by asking them) you get optimum results by keeping the Dolby of your own recorder switched off. After all, if the station provides a "Dolbyized" signal, you don't have to add your own Dolby to it. Then, in playback, you switch your Dolby back on to compensate for the high-frequency emphasis in the broadcast. For recording FM stations that don't broadcast "in Dolby," keep your own Dolby on for both recording and playback. If you find that the playback of your off-the-air recordings lacks highs and sounds dull with the Dolby on, you can either switch it off or compensate for the dullness by turning up the treble control on your receiver or amplifier.

Improving Reception

Everything said so far assumes that the FM station you are recording is coming in loud and clear. This may not always be the case. If you live in a fringe reception area or in some other location where the terrain makes FM reception difficult, your FM may be marred by background noise and distortion.

The quality of the signal you get in a given location depends partly on the sensitivity of your receiver—the more sensitive the receiver, the better it is able to pull in weak or distant stations. Even in city locations, a superior receiver tends to be more effective because it usually has better interference rejection to filter out the electrical disurbances that often impair reception in urban settings. Yet in most instances, a great improvement can be obtained simply by installing a better antenna. After all, even the best tuner is no better than the antenna to which it is connected, though a more sensitive tuner (or receiver) manages to extract better sound even from a poor antenna. An adequate antenna, therefore, is one of the basic requirements for successful off-the-air recording.

This raises the question of what constitutes adequacy in antennas. Generally speaking, a satisfactory antenna is one that 1. brings in all—or at least most—of the FM stations in your area, and 2. brings them in clear and free of background noise. The particular type of antenna to meet these demands varies according to your location.

In or near a city you are usually close to the transmitter. In that case, an indoor antenna often suffices. Out in the suburbs, a roof antenna may be advisable for best results. And in fringe areas—more than forty miles from the station—a multi-element Yagi antenna (named for its Japanese inventor) is usually indispensable for optimum FM sound. Since Yagi antennas are sensitive in one direction only, they have to be aimed at the station to be received. If the various stations you like to hear lie in different directions from your home, you may need an antenna rotor, similar to those used for TV in fringe areas.

Although the main use for Yagi antennas is in fringe areas, even in the city they offer certain advantages. City reception is often plagued by so-called multipath distortion, which results from tall, steel-frame buildings reflecting the radio signals like so many mirrors. This produces blurry sound, which is the audio equivalent of ghost images in TV. Since the Yagi antenna picks up signals from one direction only, it can be aimed at the transmitter to pick up the direct signal while rejecting the images reflected from other points.

You can get antennas that work for both the TV and FM bands. But best results are usually attained with antennas designed especially for FM, and it may be worthwhile for you to install such an antenna. Companies such as Jerrold, Winegard, and Finco offer a variety of models.

If you live in an apartment building with a master antenna, you can try connecting your FM tuner to the antenna outlet by means of an FM-TV coupler. This is a small, inexpensive device—costing only a few dollars—which lets you connect an FM set and a TV receiver to the same antenna without mutual interference. This may improve FM reception in some cases, but not necessarily. Not all TV antennas are sensitive in the

FM band; in fact, some of them are specifically designed to exclude the FM band.

Assuming that after all reasonable steps have been taken to improve FM reception, the station you want to record still comes in with a high level of background noise. In that case, you have one last resort: flip the MONO/STEREO switch on your receiver to mono. Chances are the background noise will disappear, because mono requires far less signal-strength than stereo for noise-free reception. Of course, you lose the stereo effect that way. But, under the circumstances, this may still be preferable, a clear mono signal being far more pleasant to hear and to record than a stereo signal marred by noise and distortion.

Fortunately, the difficulties described here are exceptional. The majority of listeners with suitable receivers and antennas have little trouble in collecting their radio bonanza on their tape machines.

CHAPTER 8

THE PERMANENT RECORD

Buying a record usually means you like the music so much that you want to have it for keeps. But how long will it really last? And how good will it sound after the first few plays? Granted, a pampered record, carefully dusted off before each playing and played only on component-type turntables with a lightweight tone-arm will sound almost as lush after two hundred playings as after the first. But few records get such fancy treatment. Under more casual playing conditions, an LP record often literally bites the dust after only twenty spins or so. True enough, you can still hear the music, but if the record has been allowed to gather dust and fingerprints, chances are that the surfaces crackles and the highs sound raspy. To complicate matters, the better your equipment, the worse it sounds with worn records; for components capable of coaxing the subtlest wisp of music from the record grooves will render every nick and scratch just as faithfully.

One form of life insurance for your record collection is simply to dub the disks onto tape while they are still brand new. What you then have is a "permanent record"—for tapes hardly ever wear out. They can deliver thousands of repeated playings without sounding any the worse. The original disk recording can then be preserved in prime condition as a reference from which other tape copies can be made when needed.

Why Dub?

Of course, you might ask, why not just buy prerecorded tape in the first place? Why not, indeed? But there are several reasons that make disk-to-tape dubbing preferable in many cases.

For one thing, the music you want may not be available on reel or cassette. Although the repertoire on prerecorded tapes is growing by leaps and bounds, the phonograph record still remains the main medium for recorded music, offering the widest musical choice. Particularly the more interesting items of music off the beaten path—in the classical, jazz, or folk areas —never find their way into the tape catalogues, and the record remains the main road for musical discovery. What's more, for economic reasons, even phonograph records of this type of unusual music often go out of print and become collectors' items just a short time after their release. If you have been collecting records over a period of years, chances are that you have many items no longer available, and they should certainly be preserved on the lasting and damage-resistant medium of tape.

You may also want to share your records—particularly the rare, unavailable ones—with your friends, and there is no better way than making tape copies for them. I belong to an informal tape-exchange club whose members make tape copies of their prize disks available to other members on request. That way we all benefit from each other's record-hunting instincts and plain good luck, which has enabled some of us to track down rare gems of the recorded repertoire.

Personally, I copy records onto tape for yet another reason. Often I don't like the sequence of musical numbers on an LP record. Some numbers I like better than others, and when I copy I simply skip the ones that bore me. My collection of Judy Collins and Joan Baez has become much more interesting thanks to such "censorship." Besides, many performers (or their record producers) apparently feel that they must constantly change the tempo of a program to hold the attention of their audience. That's why they usually follow a slow number with an up-beat number, and vice versa. This may be good policy in a live concert, but I find it disturbing in listening at home, for it constantly breaks the mood. That's why I rearrange the program sequence when I copy records on tape, regrouping the slower and faster numbers to avoid an emotional jolt at the start of each new selection.

I even applied this approach to some classical works. For example, I made a tape containing all the slow movements of all four Brahms symphonies, one after the other. The emotional effect is marvelous—a gradual intensification of a deeply contemplative mood—something no concertgoer or record listener who hears these works in their normal context would ever experience. On another occasion I juxtaposed the jauntier parts of Haydn symphonies with some Sousa marches. Though the effect may shock the purists, I think it's hilarious and musically very effective. Such "programming" on tape—making sound collages of similar or starkly contrasting selections—is one of the most rewarding and creative aspects of taping and will open new perspectives on music for you.

Many people also dub their own cassettes from disks simply because the sound quality of commercially prerecorded cassettes is sometimes quite poor, notably lacking in highs. This is caused by the rapid mass-duplicating process employed in making prerecorded cassettes. In the interest of lowering the manufacturer's cost, the tape is run through the duplicator so fast that the high frequencies sometimes do not properly register. That's why cassettes made at regular speed on your own machine are likely to sound a lot better than the commercially prerecorded cassettes.

I also find that cassette copies of my record collection are very handy at vacation time when I head for my hilltop cabin in the Berkshire Hills of Massachusetts. For one thing, cassettes are a lot lighter and less bulky than records, and therefore much easier to take along on trips. Besides, before I started dubbing, I often ran into trouble when lugging records to the country. Too often have I found one of my treasured disks looking limp and wilted after a journey in the hot car trunk. I have no qualms about taking cassettes with me, for they are fairly heatproof and otherwise more resistant to the perils of a nomadic life.

Record Hygiene

To get the best possible dub from a disk, the record must be perfectly clean. Before you play the record, dust it off with a special record cleaning brush obtainable at nearly all better audio shops. Such brushes—sold under such trade names as Disc-Preener and Parastat—have a special velvet covering whose nap is especially designed to reach down to the bottom of the record groove and remove deep-seated dust particles. Stay clear of the so-called antistatic sprays widely promoted for record care. Many of them leave a residual film on the record groove, which turns gummy with age and makes it more difficult for the phonograph stylus to follow the groove contours accurately. In some cases, it may even cause the tone arm to skip grooves.

My own favorite form of record hygiene—especially when dealing with neglected, dirt-encrusted disks that have smudges or fingerprints on them—is simply to wash them with lukewarm water and a mild detergent, forcing the detergent down into the grooves by gentle pressure with a foam-rubber sponge. Then I rinse the records with cold water and stand them up to dry in the disk-rack. My friends are taken aback to see me treat records like a pile of dirty dishes, but it invariably makes them sound a lot better. Of course, if you keep your records in good shape to start with—always dusting them before each spin, keeping them in their plastic dust jackets at all other times, and never touching the grooves with your fingers—such radical measures as the record laundry I have just described may not be necessary.

Checking Your Turntable

Before you start dubbing, make sure that your record player is working at its best, eliciting all the tonal nuances from the disk. After all, your tape transfer will be no better than the signal provided by your turntable or record changer. Often the phonograph stylus (or "needle," as it used to be called) has a small dust-wad on it, which prevents it from tracking the record groove properly. Such dust accumulations come as naturally to a phono stylus as dirt comes to a pig, and for the same reason. The stylus just wallows in it. During the play of a single LP side, the stylus literally sweeps up about 2½ miles of groove—the curviest, nookiest dust-catcher you ever saw. The dirt then mounts in miniature heaps on the stylus and tends to derail it from its curvy path. To remove those little dust balls, use a fine brush. Special brushes for this purpose are available in audio stores. Do this very gently, for the delicate stylus shank—engineered to be responsive to the subtlest deflections—is easily bent by rough handling. Above all, don't ever drag your finger across the stylus tip from side to side. This is invariably fatal to a modern stereo cartridge. If you must use your fingers to get obstinately encrusted dust off the stylus, move your finger across the stylus tip from back to front—in line with the stylus shank—so that it won't bend. Better yet, pry the encrusted dirt off very gently with a pin.

While we are on the subject of the phono stylus, if you believe in what used to be called "permanent needles" you might as well believe in fairy tales. Even a diamond stylus gradually wears down, and after about 2,000 playing hours you should take it to an audio dealer to have it inspected under a microscope. A worn stylus not only produces fuzzy sound; it acts in the record groove like a plowshare in its furrow and will literally ream out the music from your records.

The stylus should be suited to the type of record you want to dub. Normally this is no problem, for the elliptical styli provided on most of the better stereo cartridges today will play any kind of LP record, both mono and stereo. But if you have some old 78-rpm records you want to transfer to tape, you need a special stylus to fit into the wider grooves of those old disks. Playing a 78-rpm record with a modern LP stylus gives very poor sound and lots of background noise. Fortunately, a few manufacturers of phono equipment—notably Pickering, Stanton, Shure, and the Danish firm of Bang & Olafsen—still make special diamond tips for playing older records. These tips are interchangeable with the modern stereo styli in many of the cartridge models made by those companies, so inserting the stylus for those older records can be done quickly and easily.

Chances are that by playing your old records with these special styli, you'll get far better sound than you ever heard from them before. Some years ago a friend of mine discovered a box in his attic containing a superb batch of old swing-era recordings—Benny Goodman, Tommy Dorsey, Duke Ellington, and all the rest—which a member of his family had collected and long forgotten. After putting them through the record laundry, we were able to get remarkably fine sound from those old disks when we transferred them to tape by means of a special 78-rpm stylus. After dubbing your 78s, don't forget to put your regular stereo stylus back into your phonograph, for playing a modern record with the old-type stylus will cause permanent damage.

The most important single factor in getting optimum sound from a record player is the adjustment of the tone-arm tracking force. This is the amount of pressure exerted by the stylus on the disk. Most people believe that the lighter the tracking force, the better. They know that if the downward pressure is too heavy, it may damage the stylus, shorten the life of the records, and lessen the separation between stereo channels. But this doesn't mean that you should set your tone-arm balance for the lightest possible pressure at which your phono-cartridge will track without actually jumping out of the groove. For if the tone arm rides too light, the stylus tip does not maintain constant contact with the groove wall in loud passages, causing the stylus to make tiny jumps rather than accurately following the musical wave-forms. The result is fuzziness and distortion just when there should be a thrilling musical climax.

The manufacturers of phono cartridges usually indicate a range of recommended tracking force for each model. In my experience, the best results are usually obtained in the heavier third of that range, though this does not hold true in all cases. A professional-type tone arm with precision bearings—and hence minimum friction drag—may very well tolerate lighter tracking pressures. A little experimentation will help you find the optimum setting—usually adjustable on the tone arm by means of a sliding counterweight or a spring adjustment. At

A high-quality record player, such as this Pioneer model, is essential for optimum sound-transfer from record to tape.

any rate, don't hesitate to set the tracking force a little higher than the midpoint of the indicated range if the loud passages sound clearer that way; also, do not rely on the calibrated scale provided on most tone arms for setting the tracking force. Some of these built-in scales are accurate, others not. To be sure, doublecheck with a separate stylus-pressure gauge obtainable at audio shops for a nominal amount.

Proper maintenance of the phono cartridge and stylus assures clean sound in record playback. The tone-arm, too, must be accurately balanced. (*Courtesy Shure Bros., Inc.*)

Taping Hints

Whether you dub to reel or cassette, the procedure from this point on is the same as for recording off the air, and most of the hints given in the preceding chapters also apply here. Again, set the recording level so that the meter hits zero at the loud passages on the record. You can usually spot the loud passages simply by looking at the grooves. The loud parts look silvery, because the heavy swings of the record groove reflect more light, while the softer passages are black. So, before you actually start recording, place the tone arm on one of these silvery-looking parts of the record and set the level so that the pointer just hits zero. Presetting the level in this way has two advantages. Once it is set, you can usually let the whole record play through onto the tape without watching the meter. Also, if you are dubbing several different records onto the same tape, the loudness level on the tape will come out fairly consistent if you adjust it for each record in this manner. This, in turn, makes for relaxed listening when you play your tape, and you won't have to jump up and readjust the volume for different "takes" on the tape.

If you are recording an old monophonic record, be sure that the MONO/STEREO switch on your amplifier or receiver is in the mono position, which considerably reduces the surface noise from the older mono records.

The moment when you set down the phonograph stylus on the disk, no matter how gently you do it, usually makes a rather objectionable thumping noise. To prevent this noise from getting on the tape, I usually preset all the controls, put the machine in the RECORD mode, and then push down the PAUSE lever. Next I put the stylus on the record. Immediately afterward—before the music starts—I release the PAUSE control so that the recording starts just *after* the initial thump.

As for the scratchiness of old records, there are two schools of thought on how to deal with this problem. Some people use the scratch filter on their amplifiers or receivers to reduce the background noise. The trouble with this method is that the filter cuts out some of the high frequencies of the music along with the surface noise. Granted, many old records do not have much high-frequency content in their grooves, so you don't lose much by employing the filter. Personally, I prefer not to use the scratch filter. I believe in getting all the sound out of the record and onto the tape—scratch and all. If the noise is too troublesome in playback, I can always make adjustments later, either by switching in the scratch filter or by turning down the treble control.

Tape Speeds for Dubbing

If you work with an open-reel recorder, you have the option of selecting different tape speeds. Since old 78-rpm records rarely contain frequencies above 7,000 Hz, there is little point in using the fast 7½ ips speed. It only wastes tape since all the sonic content of these older records is easily captured at 3¾ ips. If you have a good tape recorder whose frequency range extends to 12,000 Hz and beyond even at the slower speed, you may find that most LP dubs sound quite acceptable at 3¾ ips. On the most recent records, however, the frequency response extends even beyond 15,000 Hz and you may need the 7½ ips speed to capture all the sonic brilliance of such outstanding recordings.

In any case, you can experiment and then choose the speed which gives you the least perceptible difference between the disk original and the tape copy. If you have a three-head recorder with instant monitor facilities, you can simply switch back and forth between the tape and the disk during the transfer to get an instant check on how closely the copy resembles the sound of the original. Unlike taping off the air, copying records always allows you a second chance. If anything goes wrong during the first try, you can profit by your mistakes and do it over.

Indexing

If you put many different selections on the same tape, you need some guide to find a particular piece you may want to hear. On open-reel tape you may splice in so-called leader tape, which is a strip of differently colored tape that stands out visually, so you can find a particular segment of the program just by looking at the reel. This is the method generally employed in studios. However, there are drawbacks. For one thing, you cannot record on leader tape. It serves merely for identification. Hence you can record in one direction only, since the leader tape between selections breaks up the continuity of the recording surface. Besides, splicing in the leader tape is a cumbersome task that most amateurs just won't bother with. Moreover, on cassettes—the most popular of all tape formats—there is no possibility of splicing at all.

That's why most hobbyists rely on the tape counter on their machines to provide indexing of selections on a given reel or cassette. At the beginning of the cassette or reel, just set the counter to zero. Then write down the number on the counter at the beginning of each new selection dubbed on the tape, along with the name of the piece and the artist. Put this information on an index card and keep the card inside the box of your reel or cassette. Simply by referring to the counter numbers, you can quickly locate the selection you want. Just set the counter at zero at the start of each tape.

Making Your Tapes Last

At the outset of this chapter I stressed the fact that, by dubbing your disks, you make the record permanent. This raises the question: how long do tapes last?

Nobody really knows, but this much is certain: The material from which modern recording tapes are made—polyester, generally known by the trade name Mylar—is one of the most stable substances known. As a medium for information storage, tape is more durable than engraved stone, being less subject to erosion. In fact, with a growing segment of human knowledge being entrusted to tape in the form of computerized data banks and information libraries, tape makers have been paying attention to what they call "archival properties"—the ability of tape-recorded information to withstand the ravages of time.

Most amateur tapesters may be unconcerned about leaving a historical record, but they may well wonder if a cherished musical performance will last a lifetime or if a recorded family memento may yet be enjoyed when their grandchildren are grown up. While the basic tape material will not degenerate, the longevity of your recordings depends partly on the way you treat them. Your tapes will probably outlast you if you observe the following precautions:

1. Rewind tapes *before* each playing, not after. The fast rewind speed puts too much tension on the tape for extended storage. The higher tension not only may stretch the tape, it also makes the separate tape layers stick more closely to each other. This may cause "print-through," by which the magnetic traces on one turn of the tape on the reel (or within the cassette) imprint themselves on the adjoining turn. This is a very slow process, but over the years it may cause strange echo effects on the tape. Tapes should therefore be stored on the takeup reel after they have been wound slowly and evenly during recording or play.

2. Inspect open-reel tape for even winding. Check the tape for edges sticking out of the pack. This will cause playback difficulties if the tape is left for a long time this way. Rewinding the tape slowly, at regular playing speed, solves this problem.

3. Keep the tape dust-free. Immediately after use, put the reels or cassettes back in their boxes.

4. Keep the tape recorder clean. Dust and oxide deposits shaved off the tape from previous playings can abrade the tape. Cleaning the tape path through the recorder with a soft

cloth removes such particles. Also, be sure the machine doesn't jerk the tape when starting and stopping. This can stretch the tape, causing pitch variations.

5. Keep tapes away from heat and direct sunlight. Avoid hot attics or damp basements. Also be careful never to expose tape to magnetic fields surrounding such devices as loudspeakers or electric motors and transformers. Even ordinary AC lamp cord can cause magnetic trouble if the tapes are left near the cord for long periods.

While most modern tapes are made of polyester, you may have some older recordings still made on acetate tape. (You can tell the difference by holding up the tape reel against the light. Polyester is opaque, acetate is translucent.) If you find that some of your older reels are acetate, it may be a good idea to rerecord them on polyester. The reason for this is that acetate grows brittle and becomes virtually unplayable after about twenty years. Of course, you'll need two tape recorders to do this—one for playback and one for recording. Possibly you can borrow an extra recorder for such transfers.

Keeping Out of Jail

Whenever you copy a phonograph record, you may be infringing upon copyright law. The same is true if you record off the air, although some courts have held that anything broadcast thereby becomes public domain. Composers of broadcast music and authors of broadcast plays or telecast films have naturally contested such rulings, and in general this remains a gray area where legal and moral rights and wrongs are a bit fuzzy.

As for the moral aspect of taping off the air or copying records, the American philosopher Craig Stark has devoted a learned and delightful essay to the question of whether a tape recorder hooked up to a stereo sound system is simply "a license to steal."

Just about everyone, argues Professor Stark, is turning his

tape recorder into "a kit of burglar's tools," and, as usual, the common man justifies such larceny by the convenient principle that if everybody does it, it must be all right. Such attitudes, the philosopher observes, are often rationalized on the grounds that "American society itself is based on the principle of the 'rip-off' since—as another writer puts it—'our most respected citizens are the businessmen who have most successfully held up most of the people.' One reader once wrote to me, after I had raised these questions in a magazine article, that 'If God didn't want us to dub recordings, he wouldn't have given us tape recorders.'"

Professor Stark believes it is clearly immoral to appropriate an artistic product while doing "nothing for the artists who create the aesthetic values of the recordings we enjoy." He points out that as many as 87 percent of all classical recordings fail to break even, and when income from record sales is further reduced by tape dubbing, the continued availability of good music on records is clearly endangered. Of course, in the case of out-of-print records, which cannot be bought commercially, nobody loses potential sales through dubbing and nobody is directly harmed. With erudite references to ethical theorists from Aristotle to Kant, Professor Stark concludes that in that particular case, dubbing is morally permissible.

My own sensibilities are not as finely tuned, and my main concern is just to stay out of jail. This simplifies matters. A few basic rules will help avoid legal entanglement. Above all, never record anything from the air channels used for police communications. The very fact that you possess such recordings may possibly get you into trouble if it becomes known, for there are rather strict laws concerning the monitoring of such broadcasts. As for general broadcasts and dubbing records, you'll be on safe ground as long as you keep your recording activities a strictly private affair. Don't distribute your dubbings beyond your immediate circle of friends and, above all, don't sell them for money.

If you want to produce tapes for wider distribution, obtain

written permission from the following, wherever applicable: 1. the person whose performance or voice you have recorded; 2. the company or agency to which the recorded performer may be under exclusive contract; 3. the copyright owner of the subject matter of the performance, whether it be a piece of music or a verbal script; 4. the broadcaster of the program you have taped off the air or the manufacturer of the record you have dubbed.

Or, as any lawyer will advise you, if you can't be good, be careful.

CHAPTER 9

RECORDING LIVE

The best-tasting fish are the ones you catch yourself. It's the same way with recordings. The most treasured items in your tape collection will be those you "captured" yourself; for those recordings are, in a very real sense, parts of your own life.

There may be a concert at the high school with your boy playing second trombone. Or maybe your wife's sister is finally getting married, and you want to present the young couple with a sonic memento of the ceremony. Or you are practicing that speech you are going to give at the convention of Midwest Marketing Executives. Or there's a thrush singing in the shrubbery. The occasions are countless.

There is something immensely exciting about holding fast the living voices of your friends and your family. I, for one, find popping up a mike as exhilarating as breaking out a bottle of champagne, and making a successful recording of a personal event always seems cause for celebration.

Of course, perfection in recording, as in other things, is largely a matter of practice and experience. You may not be able to equal the professionals on your first try, or maybe not even on your fifteenth. But by following some of the basic suggestions given in this chapter, you will be able to come up with recordings that are a pleasure to hear again and again.

The simplest recording situation is to record the speaking voice of a single person. A single mike and monophonic recording usually suffice for this purpose. The only precaution is to avoid placing the mike too close to the mouth of the speaker. Allow a distance of at least eight to ten inches, and if you find that the recording sounds too sibilant—with undue emphasis on such consonants as *p* and *s*—move the mike at a slight angle from the speaker rather than placing it directly in front of him or her.

Even for casual recording at home—except for simple speech—I usually set up two mikes for stereo. Particularly if recording at a party or a gathering of any kind, I find that the stereo effect helps to sort out the individual voices that might otherwise overlap and obscure each other. Besides, working with two mikes often results in a sense of realism and natural balance not easily achieved with a single mike.

We have already talked about microphones and their characteristics in an earlier chapter. Now we must consider the matter of microphone placement. The small desk stands furnished with some mikes do not allow sufficient flexibility for really effective placement. You will need a proper floor stand for each microphone. Even an inexpensive stand, costing about $15–20, will do very well for most purposes. The fancier and more expensive mike stands have built-in shock absorbers so that footsteps and other floor vibrations will not reach the mike and cause rumbling noises on the tape. But you can filter out floor vibrations simply by putting a foam-rubber mat or a thick typewriter pad under the mike stand. If you want to position the mike very precisely, you may want to add a boom attachment to your mike stand, which lets you swing the mike sideways and permits additional height adjustment. Such boom attachments are available at specialized audio shops, and the cheapest ones cost about $15.

If you want to do professional-sounding recordings of orchestras or rock or jazz groups, it is often helpful to use more than two mikes to highlight individual instruments or blend in a soloist with an orchestral background. In that case, you need a so-called microphone mixer which lets you connect extra mikes for each stereo channel and adjust their relative balance. Such mixers vary in price according to complexity (the number of mixes that can be connected) but the least expensive ones, accepting four separate microphones, cost between $20 and $40.

Before going afield with your tape recorder to tape local concerts or other public events, it's a good idea to gain some practice by recording at home. Family life is full of events to be documented on tape. The growing up of your children (taping their conversation over a period of years), the unfolding of their musical or dramatic talents, their progress in the knowledge of a second language, their interaction with friends at birthday parties and similar occasions, their audible delight at the unwrapping of Christmas presents—all these are inexhaustible "source materials" for the tape hobbyist.

Sociologists tell us that modern technology disrupts family life by making too many diversions too easily accessible. Yet a tape recorder is a technical innovation with exactly the opposite effect. It can serve as a center of interest that draws a family together.

A friend of mine, for example, has revived the old custom of reading aloud. At present, he and his family are putting on their own drama festival, taping plays from Marlowe to Miller with each family member taking one of the roles. They act out the parts, moving about as if on stage, and the stereo effect suggests the stage action in playback. If the available boy/girl ratio doesn't match the cast of characters in a particular play, friends and acquaintances usually provide eager extras for the tape theater. My friend also has several sound-effects records so that he can later blend in the appropriate sonic background by sound-on-sound recording, along with suitable music to suggest scene changes. The whole production often sounds like professional radio drama.

If you or anyone else in your family sings or plays an instrument, live recording is a tremendous teaching tool; for it allows us—to paraphrase Robert Burns—to hear ourselves as others hear us, which is sometimes sobering but always instructive. Particularly for playing in groups or combos, the recorder is an excellent training aid, for it lets the players unravel the muddles where they sometimes get lost. I'm always delighted to hear the reactions of combo players listening to the playback: "Gee, that was *me* messing up!" or "Hey, you're coming in a beat early!"

Tips and Techniques

Probably the most important thing to keep in mind for setting up mikes is that, like photography, sound recording is a highly artificial process. Just as camera lenses don't "see" in the same ways that human eyes do, microphones don't "hear" in the same ways human ears do. That's why it takes a certain amount of skill and experience in mike placement to come up with recordings that sound really natural. After a dozen tries or so, you are bound to get the feel of it. But for starters, let's run down some basic mike-placement patterns for the most typical recording situations.

Vocal with Guitar. Place one mike about eight inches from the singer's mouth, slightly at an angle to the line of voice projection. Put the other mike in a lower position so that it is on a level with the sound hole of the guitar at a distance of about eight inches. If the guitar sound is too "plucky"—i.e., if it is picking up too much finger noise from the strings—move the mike a couple of inches further away.

Solo Piano. Open the piano top and place one mike about six to eight feet away from the piano approximately in line with the strut that holds up the piano lid. Place the other mike closer, almost directly over the piano string board, but turn down the gain on this channel, otherwise the piano will sound hard and clangy. If clangy sound persists, move the second mike further away from the string board. If more room echo is desired, move the first mike still further away from the piano.

For recording an upright piano or spinet, a microphone distance of six to eight feet usually works out well.

Singer Accompanying Himself on the Piano. Leave the piano lid closed and place the first mike squarely in front of the singer at a distance of about ten inches. A mike boom is helpful in this case to swing the mike across the piano case. The other mike should be placed at a distance of about two feet in line with the crook in the piano case.

String Quartet. With the four players facing each other, raise the mikes to a level slightly above their heads. If your mike stands are not tall enough, place them on chairs. One mike should be located between the first violin and viola, the other between second violin and cello. This arrangement provides good stereo localization of the four players. Many recordists, however, prefer another pickup pattern which does not pinpoint the location of the players but provides smoother blending of the ensemble and greater ambience effect. To accomplish this, place one mike at the center of the group (again raising it to a level about their heads) and set up the other mike in front of the group at a distance between six and ten feet, depending on how much room echo you want. As always, echo increases with mike distance.

Folk or Country Vocal Group. Place the two mikes four to six feet apart in front of the group, pointing them slightly away from each other aiming at the two outermost members of the group. If one of the singers does not register clearly, move him or her slightly forward, closer to the mikes. If, on the other hand, one of the group predominates, move him toward the back. The overall blend of vocal groups can usually be greatly improved by not lining them up parallel to a wall of the room but diagonally across one of the corners. Never place a group in the middle of the room, as this will usually result in spotty and ragged sound.

Rock or Pop Group. If working with only two mikes, follow the same procedure as for folk and country vocal groups. Sometimes you can improve results by hanging the mikes from the ceiling directly over the group, with the mikes spaced about four to six feet apart.

If you have extra mikes and a mixer, you can get an added feeling of impact and "bite" by placing so-called accent mikes directly in front of the drums and other instruments you want to bring out more dramatically. The accent mikes should be mixed in at a lower level than the two main mikes for the overall pickup.

Orchestra or Choir. The higher the mikes, the better the blend. They should be above the heads of the musicians, so put the mike stands on chairs or suspend the mikes from the ceiling. Put both mikes at least fifteen feet in front of the first row of singers or players. Increase this distance if more room echo is wanted. Space the two mikes so that they divide the width of the stage into three equal parts.

If the mikes are too close, musicians or singers in the front row will overbalance the rest. If you're recording a public concert, the best procedure is to make a test recording during rehearsal and then adjust the mike placement after sampling the results. If you have no chance to make a trial run, it's always safer to place the mikes at a greater distance. You may lose some definition of individual instruments, but the overall blend and hence the general musical impression will be more agreeable.

Don'ts and Other Precautions

A handful of "don'ts" may help you avoid some of the most common beginner's mistakes.

–Don't move in too close. The usual temptation is to place the mikes too close to the performers. This often results in overloading, distortion, and generally unbalanced pickup. As a rough guide for miking various instruments, keep these minimum distances: 15 to 35 inches for strings and woodwinds or harmonica; at least 6½ feet for brass instruments; at least 25 to 35 inches for a vocalist who projects in full voice. With folk singers, who rarely use projective vocal styles, you can move in closer.

—Don't hand-hold a mike while recording. Even slight motions of your hand will alter the balance. The exception to this rule is casual interview recording with portable machines, where you have to catch people on the run. If you have to hold the mike in your hand in such situations, try to keep a fairly constant distance between mike and speaker.

—Don't put a mike on the same table with the recorder, for the mike will pick up vibrations from the mechanism. Also, it will record the clicks as you operate the switches and controls.

—Don't put the mike on a piano. It will pick up the mechanical noise of the keys and the pedals through the frame of the instrument. Use the piano-recording technique described earlier.

—Don't put a mike in the middle of the room. Sound reflections from all the walls converge at the center of the room and cancel each other out through phase interference. The result will be loss of bass and generally ragged sound.

The acoustics of the room or hall in which you are recording naturally affect the tonal character of the recording. If you are recording in your own home, you can to some extent modify the acoustics to get better results. If your living room produces too much echo, lay some scatter rugs over bare floor areas and draw the curtains across large windows, which often produce very hard-sounding reflections. It also helps to break up the sound by placing hangings or installing shelves on large surfaces of naked plaster walls. If, on the other hand, your room is too sound-absorbent, do the reverse: remove rugs, upholstered chairs, and pillows before a recording session and draw back the curtains from the windows.

If you are recording in an unfamiliar location, you can

quickly check its acoustic properties simply by clapping your hands. If the clap sounds hollow, the room is too reverberant. You can counteract this by moving the mikes a little closer to the performers than under normal circumstances. If the clap of your hands sounds dull and smothered, the room is acoustically "dead." In that case increase the mike distance from the performers.

On Location

As you gain a reputation as a recordist among your friends, you will probably be asked to do occasional on-location jobs, such as recording church services, school concerts, and the like. In that case, be sure to arrive at the scene well ahead of the scheduled time for the event to set up your equipment. In most cases, you will be using your regular tape deck; for as long as you have access to an electric outlet, there is no need for special battery-powered equipment. But be sure to bring plenty of electrical extension cord so you can reach the nearest plug. Then set up the deck at an inobtrusive location, preferably out of sight of the audience. Of course, you must also make sure that your microphone cables are long enough. You can buy microphone extension cables in 100-foot rolls for about $7–10 per roll. If, as is most likely, you are using low-impedance microphones, specify low-impedance cables.

During the recording, monitor your "take" with earphones. The best kind for this purpose are earphones with padding to shut out all surrounding sounds. With them you will hear exactly what your mikes are "hearing" so you have an instant check on the effectiveness of your mike placement. Most better tape decks have a plug-in jack for earphones. Most modern equipment is designed for low-impedance earphones; look in the specifications for your deck to find out whether your machine takes high- or low-impedance earphones. While you listen, make sure that each ear doesn't seem to hear the music "separately" from the other ear. If this occurs, it is a sign of

A pair of stereo earphones to let you monitor an on-location recording while you make it. (*Courtesy Koss Corp.*)

excessive stereo separation and that the mikes are too far apart. This happens more often with cardioid microphones, for their sound-acceptance pattern falls off sharply for sounds incident at an angle of more than 45 degrees from the direction in which the mike is pointing.

My own main experience in on-location recording has been at church weddings, which represent a rather difficult recording task because of the many sound sources involved. I have usually solved the problem by putting one omnidirectional mike slightly in front of and to the side of the bride. The reason for putting the mike on the bride's side is that she usually speaks the fateful words in a softer voice than the groom. Since the mike is omnidirectional, it will also pick up the voices of the groom and the minister. The second mike I place farther up front, near the altar, where it picks up the organ and choir while being far enough from the audience to minimize distracting noise. Of course, since loudness levels during the ceremony are usually quite unpredictable, you'll have to watch the meters closely and adjust the gain accordingly.

An input mixing panel is useful if you are blending the signals from many microphones. This console has controls for eight separate microphones, plus facilities for creating special sound effects. (*Courtesy Shure Bros., Inc.*)

The best cooks rarely follow a recipe literally. Similarly, let me emphasize once more that the suggestions given here need not be followed rigidly. Rather, they are "starting positions" which you may modify to alter the sound to your own taste. In fact, I should like to encourage experimentation of this type. Since most amateur tapesters lack the kind of sound mixing facilities found in professional studios, variations of mike placement are probably your most effective means of controlling the kind of sound you get on your recordings.

CHAPTER 10

OFFBEAT AND ON-THE-GO

The only limit to the variety of pleasure in taping is your own imagination. So, to crank up your originality and inventiveness, let me briefly tell you about some of my own somewhat offbeat recording projects. Most of these involve small, portable equipment—but we'll take up the technicalities later.

Tape Cookery

Tape cookery, for example, is one of our regular kitchen routines. The other day when my wife put together her special French veal stew again, I just popped a little cassette machine right next to her on the kitchen counter. And since she is an easy and informative talker, I just asked her to give a running description of everything she did to the stew, along with whatever other commentary came to mind. Then we ate the stew and sent the cassette to a friend in Boston who is also crazy about veal.

Couldn't we have just sent the recipe? Sure, but the results might not have been the same. For a recipe, at best, is a kind of shorthand for the intricacies and nuances that make the real difference in cooking. You can, so to speak, put the substance on paper, but the flavor gets lost. The cassette, by contrast, describes every single step, every motion, and every pinch of spice in exactly the time sequence of the actual stew preparation. If you follow taped cooking instructions, you

Tape cookery for the perfect soufflé. The small portable cassette recorder next to the milk bottle replays an off-the-air tape of Julia Child.

don't have to dash back and forth to look at the recipe. You simply follow the vocal advice on what to do and when. Of course, in return for sending out our own specialties on tape, we receive similar gastronomic cassettes from our friends, all of which add up to delightful evenings at table and the gradual disappearance of my waist.

We have been so emboldened by our success at tape cookery that we sometimes invite people to dinner. For these occasions we keep an audio guestbook. We ask our visitors to record their impressions of the meal and whatever else they feel like saying. This gives us far more vivid mementos of our social life than any guest book could provide, including one concise after-dinner comment by my friend Etti: "If you can't say anything nice, don't say anything at all."

Oral History

One of my more footloose recording projects is the assembling of an oral history of Berkshire County. Up in the hills where we have our summer cabin, people live to a ripe age, and many old-timers are full of tales of long ago. Contrary to what you might suppose, most of them aren't in the least mike-shy. In fact, the presence of the recorder seems to bring out in them long-buried recollections and the kind of narrative embroidery that used to be heard on winter evenings down at the General Store before it was replaced by a supermarket seven miles up the road at the new shopping center.

The old farmers and local craftsmen who still live in the area would never keep a journal or write their memoirs, but their heads are full of anecdotes and incidents that throw a vivid light on ways of life long since lost, and the pace of their speech and the Yankee twang of their voices lends a flavor and immediacy to their accounts that the printed word could not convey. In this sense, I feel that the tape recorder is bringing back the idea of oral history which gave such incomparable vividness to the great chronicles of the distant past—such as the works of Herodotus and Homer—long before officious scribes took over the task of recording human events.

I started my local history project by asking the old-timers about their parents and their childhood neighbors, about their travels and their working lives, their occasional trips to the city, their contact with different ethnic groups and their attitudes toward them, and I was amazed at the amount of detail they remembered—informative trifles that set the character of daily life but which are rarely mentioned in historical writings. Most of all, I was impressed by the depth of informal and profoundly human understanding that came to light in their stories. We talked about the first cars they owned, about the time electricity came in, the old volunteer fire company, and how the bears used to come down from New Hampshire to scavenge near the farms.

It was the story about the bears that gave me the idea of branching out into collecting tales of hunting and fishing—and those cassettes turned out rare documents of descriptive fervor.

Doing recordings of this sort, you soon develop the interview technique of an experienced journalist. For example, you avoid asking questions that can simply be answered by a yes or no. Instead of asking, "Have you really lived here for sixty years?" you learn to ask, "What was it like when you first came here?" You soon get the feel of what it takes to draw out a particular person. You learn to listen.

When I first started my project, I got so wrapped up in my prepared questions that my own words tended to dominate the tape. Now I rarely prepare questions for an interview. Rather, I let the subject lead me, letting him or her call the turns of our talk.

Tape Maps

While we are on the subject of country exploits, I also use cassettes to help my friends find their way to my house. It isn't easy. The very privacy that makes my cabin so attractive to me also entails an approach over all kinds of twisting backroads and half a dozen easy-to-miss turnoffs. Having lost—temporarily—a number of friends that way, I finally hit on the idea of taping approach instructions in "real time" (as the computer people say); that is, giving directions paced for traveling at the average speed of 25 mph, which is about as fast as you can go in my part of the world. Since most of my friends have portable cassette machines or cassette players in their cars, this proved a useful and reliable navigation device.

I drove the stretch from the nearest highway turnoff with the portable recorder next to me in the front seat and, as I progressed, described the road and the landmarks along the way, giving warnings where to slow down for bumps or watch out for one-lane bridges. This method has been so successful that I am now making taped road guides for visitors wanting to do some sight-seeing in the area.

Speaking of tapes during car trips, I also found taped storytelling a great help for keeping children reasonably quiet in the back of the station wagon on longer journeys. The younger kids usually like to hear the same story over again and again, and tape provides a natural method for such repetition.

Inventories

But not all offbeat taping is strictly for fun. Some of it is strictly utilitarian. At the suggestion of my insurance agent I taped a complete household inventory. To do this with pencil and paper would be a very awkward and time-consuming procedure. Yet with a small cassette machine in your hand you can just walk around each room in your house and simply name everything you see. Don't attempt to organize or classify your possessions. Just make a verbal note of each as you come to it.

For insurance purposes, you should describe every piece of furniture or other item as to kind, finish, color, and purchase price, if you remember it. Just keep on talking as long as you can think of anything to say. That is the kind of information an expert appraiser would provide. If you should lose your belongings through fire or some other misfortune, the tape will give your insurance company a reliable guide as to the extent of your loss. It may also help you in replacing your possessions as far as that is possible. Of course, this cassette should not be kept in your own house. Either put it in your safe-deposit box or in some other separate and secure location and hope that this is the one tape you'll never have to play again.

This kind of taped inventory is also useful in business, particularly if you have to count items on shelves or in bins and need your hands for checking so that you can't write at the same time.

Strictly Business

Obviously there are innumerable other business applications for a small tape recorder. I know a Manhattan ad man who confesses "I sleep with my tape recorder. I used to wake up in the middle of the night with an idea for an ad. But I'd forget it again before I managed to get out of bed to write it down. Now, even when I'm still half asleep, I just push the button and mumble into the mike. I don't even have to get out from under the covers." I told this to a psychiatrist who now urges his patients to record their dreams in the same way before they escape them.

A Boston folk singer always carries a cassette recorder in his car. "I think of new tunes when I'm driving. I just pick up the mike and sing it right out into the machine. I can do it all with one hand."

Salesmen all over the country use cassette recorders to rehearse their pitch. And after visiting customers, they record their impressions of the sales call and mail the cassettes back to the home office. The immediacy of these voice memos conveys the nature of a prospect or a problem much more realistically than a written sales report.

Tape Goes to School

In some colleges and universities, the cassette recorder has virtually replaced the notebook. Students feel that, thanks to their recorders, they can now pay attention to the professor instead of having to concentrate on their scribbling. Later, when they listen to the playback, they can organize their notes more efficiently, freed from the pressure of having to keep up with the lecture. Retention of the material and the level of understanding is greatly improved in this way. Particularly in language classes, where repetition is the key to learning, tapes are invaluable, especially since they let you hear your own recorded pronunciation in direct comparison with that of the teacher.

Instructors often use cassettes to check their own lectures.

None of the professor's fine points are lost on the student who reviews the lecture verbatim from her tape. A battery-powered portable cassette machine makes an excellent teaching tool.

This is especially helpful to teachers who prefer to speak spontaneously rather than rely on prepared notes. Later evaluation of their spontaneous presentation allows them to spot omission of important material and improve the general organization of their lectures.

Many students also do their library work with a tape recorder. If they have to consult books that cannot be removed from the library, they no longer have to copy long passages. They simply read the material softly into the microphone. This can be done in a whisper, so as not to disturb anyone

Library research is a lot less laborious with a portable cassette recorder.

else: but the whispers emerge clear and understandable when amplified in playback. The students can then evaluate the material they have researched and select the appropriate references and quotes for their papers.

At Law

I know a lawyer who uses the same technique in his visits to the Bar Association library. In essence, he uses his recorder as a kind of oral notebook as he delves into the legal volumes. Sometimes he also uses the recorder in taking depositions from his clients, which are later transcribed by his secretary. That way, the actual words of the client are always available for reference.

This brings up the question as to the use of tape as legal evidence—a matter given much attention a while back in connection with a highly prominent tape hobbyist in Washington. Opinions vary in this complex legal area, largely because of the possibility of tape being "edited."

Apropos of the legal aspects of taping, you are by law required to inform a person at the other end of a telephone conversation if you have a recorder hooked up to the phone. But at least the technical part of the hookup is simple. For just a few dollars you can buy a telephone pickup, which is a magnetic coil sensing the voice currents in the telephone receiver for both sides of the conversation. The device attaches to the phone by means of a suction cup and the cable just plugs into the microphone input of your recorder.

Tape and the Writer

If you write for publication, you will find a portable cassette machine invaluable in preparing your articles. It is much easier to conduct an interview if you just keep your recorder running than if you are frantically trying to follow a conversation with pencil and paper—especially if you have to think about asking the right questions at the same time. Of course, this works only if the person you are interviewing is not intimidated by the recorder. Nowadays few people are. But it helps if you put your subject at ease. If possible, set up the recorder before the interview and put it discreetly out of the immediate line of sight—off near the far edge of the desk or table, or next to you on a couch. Today's better portables have special devices—the automatic level controls to be explained later—that allow them to pick up speech intelligibly even at a slight distance. What you are after, under such circumstances, is merely intelligibility—fidelity doesn't matter.

If you must set up the recorder in the presence of the person you are about to interview, refer to it in a casual, offhand way. Just say, "That's my note-taking machine—just so I don't misquote you." This usually breaks the ice.

Only once did I encounter resistance. A banker I was interviewing objected to the recorder. Bankers, after all, tend to be secretive. I looked him in the eye and explained very seriously: "But this is a very special machine. It doesn't record what you don't tell me." He grinned and the interview went fine.

Small Fry

There is a special group of people who also have to be coaxed a little to open up in front of a recorder—children. Once I got around this problem simply by bugging a Christmas tree, hiding the mike among the decorations. The result was a sonic document of Sonja and Andrea at 4½ and 6, respectively. The recording is a joy, and I regard this particular tape as conclusive evidence that this world can be a wonderful place.

On other occasions I managed to loosen up children before the mike by acting out parts of their favorite stories. By the time they become sufficiently engrossed in my antics to forget about the recorder, I suddenly look puzzled and helpless and ask pathetically: "I forgot the rest—what happens next?" I, and the tape, usually get exhaustive and emphatic answers,

Some kids are mike-shy and need persuasion to express themselves on tape. This one obviously needs no encouragement to rehearse his part as the Tin Man in *The Wizard of Oz.* (*Courtesy 3M Company*)

and from then on it is usually quite easy to capture a revealing aspect of the child's personality.

Once I took Sonja to the zoo, carrying the recorder over my shoulder like a camera and letting the tape roll as we ambled from cage to cage. The information I obtained would astound zoologists.

If you have children, you will no doubt devise your own ways of eliciting recordable traces of their ever-changing lives, and the nicest part of it is that—while the tape rolls—you too may become something of a child again.

Traveling with Tape

Sound is where you find it. You can go on tape safaris, bagging your trophies at folk-song hootenannies, at carnivals and county fairs, at impromptu jam sessions, at the initiation rites of the Navahos or snake-charming contests in Katmandu, or at evensong in Westminster Abbey. Sonic mementos from your travels, particularly when accompanied by your color slides, will bring back treasured experiences time and time again. Among my own travel "takes," I particularly treasure a Viennese folk singer encountered at the *Heurigen,* one of those tree-shaded courtyards in the Vienna Woods where the locals sample the new wine on summer evenings. Another time, at a lake near Salzburg, I caught a little brass band tootling sweetly on a barge as they sailed from village to village to present their Sunday serenade. And once a company of miners in a Welsh tavern obliged me with some magnificent choral singing when they noticed that I carried a recorder. Some of my friends brought back surprisingly good tapes of the great carillon at Bruges, gypsy fiddlers from a *czarda* near Temesvár, running commentary from a New York taxi driver, a dragon dance from a festival at Kyoto, brass fanfares from the towers of Basel, the stentorian toots of the Alphorn accompanied by cowbells from a valley in Graubünden, the swirling pipes of Highlanders trooping through Aberdeen, and a little vituperation in the local language from an undertipped Sardinian waiter. But before you roam the world in search of your own sonic adventures, we must give preparatory thought to the required hardware.

Portable Recorders

The basic tool for on-the-go recording is the small, battery-powered, portable cassette recorder. Since ultimate fidelity is rarely a major concern in this type of recording, the limitations of such equipment are not a significant drawback. I my-

A typical sonic travel memento: a Jamaican family performs on native instruments.

self roam about with a small, lightweight recorder of this kind, which runs on either batteries or AC and can also be operated with an adapter from the cigarette-lighter socket in my car. It also has a built-in mike, which is extremely convenient. Granted, the internal mike picks up a little background noise from the drive motor, so I also carry an external microphone, which I plug in for more critical recording. But otherwise the little machine does an admirable job on speaking voices and a tolerable job on music. Its main weakness for musical recordings is that it naturally has more flutter and wow than heavier recorders with larger flywheels and heftier motors.

The main shortcoming of small portables is in playback. With their tiny speakers and puny amplifying circuits, they can't possibly produce sound of adequate range and volume. That is why I use the built-in playback facilities only for instant, on-location checks on the recording. When I get back home, I bypass the playback bottleneck by playing the cassettes through my regular, home-based deck and stereo system, which shows up the surprisingly good quality of recording obtainable even with portable equipment.

When shopping for a portable, avoid the cheapest models generally sold for less than $40 in glorified junk shops and other bargain emporia. As a rule, such machines are poorly designed, flimsily constructed, and thoroughly unreliable. But portables in the $80 to $140 range are often remarkably sturdy and serviceable. Many of them have provisions for rechargeable batteries, which is a great help for on-the-spot recording. Even where power outlets are available, it is not always convenient to run a cord to them. Putting the small, self-contained recorder on a table or on a chair and letting it run off the battery is simpler and less obtrusive, and afterward you can recharge the battery simply by plugging in the recorder overnight—but watch those foreign voltages! The rechargeable battery may cost $20 or thereabouts, but it can be recharged several hundred times, which makes it more economical in the long run than buying fresh batteries for each taping project.

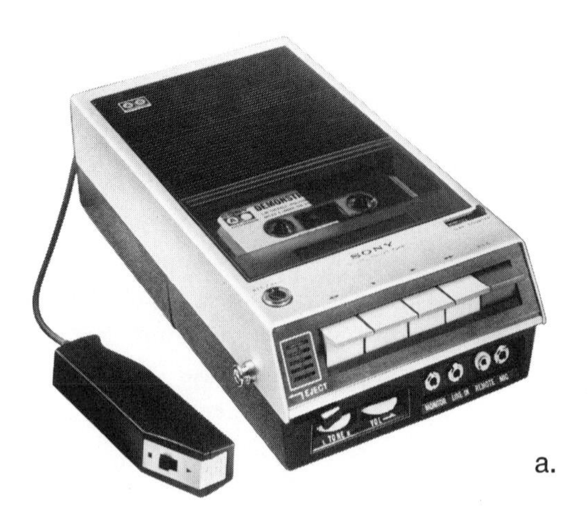

a.

b.

Portable cassette recorders make rewarding travel companions.

a. With its built-in mike, this cassette model permits unobtrusive recording in any location. At the left is an optional remote on/off switch.

b. Particularly convenient for travel is a high-quality portable radio with built-in cassette recorder. Built-in mike makes for easy on-location recording—and by taping local broadcasts you can recapture added travel memories. (*Courtesy Sony Corp.*)

Most of the better portables also have a meter indicating the battery charge. If you use regular batteries, always use the alkaline type; they last longer and provide power for several hours of continual use. Even so, it's a good idea to put in new batteries each time you go out to record a talk, conference, or whatnot. Nothing is more frustrating than discovering afterward that your battery ran down in the middle of the recording.

A highly useful feature obtainable on the better portables is the automatic level control, ALC for short. This automatically adjusts the recording level to the loudness of the incoming sound, preventing distortion on loudness peaks and making it unnecessary to watch the meter during the recording. The ALC also automatically increases the gain when the incoming sound is soft—a great advantage in recording conferences where people speak from different distances to the recorder. The ALC makes them all equally intelligible, regardless of where they sit around the conference table. Of course, the ALC action wipes out differences between loud and soft in musical recordings, thus robbing the music of its expressiveness. The better portables therefore feature a switch to cut out the ALC, if desired.

The best way to judge a portable recorder is simply to handle it, getting the feel of its weight and size. Then place it on a table, set it on RECORD and speak casually while walking about to various points of the room. Then play back the recording to check if the ALC properly compensates for the loudness differences. Take the recorded cassette home to play on your regular stereo deck to check the fidelity of the recording, paying particular attention to the level of background noise picked up internally by the built-in mike and to the clarity of speech consonants. Another useful trick is to jangle some keys near the mike and then play the cassette on a big stereo system to evaluate the realism of the sound. Key-jangling clearly reveals the high-frequency response essential to good speech intelligibility. If the clink of the keys comes through clearly, so will human speech. Some of the better portables have a frequency response up to 10,000 or 12,000 Hz, while the poorer ones cut off at around 7,000 Hz.

To check the mechanical quality of a portable recorder, load it with a type C-120 cassette, which is designed for two hours' playing time. These cassettes contain extremely thin tape so as to accommodate the extra length on their internal reels. This makes the tape particularly sensitive to stretching, snagging, or breaking on a recorder whose mechanism is incapable of maintaining fairly constant tape tension in all modes of operation. If such a tape starts, stops, rewinds, and runs in "fast forward" without any mishap—both at the beginning and in the middle of the tape—you can take it for granted that the recorder is well designed.

Top-Quality Portables

If you plan to do more serious sound-hunting and require genuine fidelity in your on-the-go recordings, inexpensive portables will not suffice. Fortunately, recent technology has produced portable recorders to compare with some of the finest standard recorders in quality. However, such machines tend to be somewhat heavier and much more expensive. The ultimate in portable recording is attained by such open-reel models as the Nagra and the Tandberg, made in Switzerland and Norway respectively. Priced upward of $1,000, these machines are the standard for super-critical applications, such as making professional motion-picture sound tracks. The advantage of the open reel, of course, is that it permits easy editing. Moreover, these top-quality portables also have provisions for synchronizing the tape with a motion-picture camera.

Unless you are doing professional film or broadcast work, you may find a high-quality cassette portable more useful and convenient. Some of these cassette portables feature sophisticated mechanisms, such as dual capstans and flywheels to

Some portable cassette recorders offer excellent fidelity such as (a) Sony's high-quality monophonic portable or (b) Uher's surprisingly compact portable stereo cassette recorder.

reduce flutter and wow despite their compactness and relatively light construction. Sony produces an excellent monophonic portable cassette recorder whose quality certainly meets broadcast standards. And if your heart is set on stereo, you have a choice of truly remarkable portables from manufacturers like Grundig, Sony, and Nakamichi with prices ranging between $270 and $500. Such equipment lets you record any event anywhere with admirable fidelity.

On-the-go recording is an exciting experience that, in a very profound sense, widens your range of perception. Most of us experience our surroundings mainly by sight. Our auditory awareness is often limited to formally confined areas, such as speech and music, and we hardly notice the fullness and variety of sounds in our environment. One of the great benefits of offbeat and on-the-go recording is that it sharpens our cognizance of hitherto unperceived dimensions of reality. We become attuned to the sounds around us and learn to perceive more of the world with our ears. This is the real adventure.

CHAPTER 11

FANCY TRICKS

Except for the last part of this chapter, which deals with creating special sound effects on tape, the advanced techniques discussed here concern mainly the owners of open-reel recorders. So, if you are among the great majority who have opted for the convenience and simplicity of cassettes, feel free to skip ahead. Yet if you have a hankering for developing professional techniques, such as editing and multi-track recording, and have therefore chosen an open-reel recorder, this chapter will suggest ways to utilize all the possibilities inherent in your equipment.

Tape Editing

On open-reel machines featuring sound-on-sound facilities, you can do so-called electronic editing. This is by far the simplest form of tape editing and has the advantage that you don't have to snip and cut. You simply dub your program material from one tape track to another, leaving out whatever parts you don't want. This is a good way, for example, to eliminate commercials from off-the-air recordings. Conversely, you can intersperse new material onto the dubbed track, such as sound effects or background music for dramatic readings. And if you happen to make a mistake in this over-dubbing process, it's no great tragedy. The original track remains intact and you can do the whole thing over again. Only after you are fully satisfied with the sound-montage you're making will you erase the original track so that you can use it over again.

A more precise kind of tape editing requires cutting and splicing, which is not inherently difficult but takes some practice to learn. Such tape splicing can be compared to splicing motion-picture film. Yet there is one important difference: you can't see what's on the tape. You have to locate the exact splicing spot by sound. That's why it helps to have the kind of recorder on which you can hear the sound when the reels are slowly moved back and forth by hand. That way you can pinpoint the exact place at which the splice is to be made. Of course, the sounds you hear as you "rock" the tape back and forth slowly may at first seem like the growling of a disgruntled bear. But after getting used to it, you will recognize words and syllables and thus be able to delete expletives and other unwanted parts of speech or music.

Tape editing need not be limited simply to deletions. It has been used actually to change the meaning of a statement, by picking out various words from a recording and then rearranging them in a different order. That way someone can be made to say in his own voice what he never meant to say at all. Even the silences on a tape are usable material for a skilled editor. Feelings of urgency and nervousness can be evoked just by shortening the natural pauses between sentences. Conversely, a sense of calm and composure is created by lengthening them.

To make an altered tape sound natural, as if the speaker had really said the things that are being put in his mouth—or rather in his tape—the editor has to remember that people breathe when they speak. He must allow for the intake of breath when he joins phrases: as he cuts and splices, he must be careful not to remove the breathing sounds. Above all, he must never accidentally join two intakes of breath in a row. To do so would make the person speaking sound as though he were swelling up like a balloon.

It is difficult to give specific instructions for tape editing. The necessary skill comes with experience. But it is possible to point out the basic tools. One needs a tape splicer, splicing tape, a marking pencil, and a single-edged razor blade. The splicer and the splicing tape can be bought at nominal cost at almost any audio shop. I prefer the simplest kind of splicing device, called a splicing block—something like the miter box used by carpenters. You just lay the tape into a groove that holds it firm and then cut across the tape at angles indicated by slots in the splicing block. For a very sharp sound division —such as between words and syllables—cut across the tape at a 90-degree angle. Otherwise cut at a 45-degree angle, which tends to produce a stronger splice. Since all the tape ends are cut at precisely matching angles, thanks to the splicing block, it is easy to join them end-to-end and connect them by putting the pressure-sensitive splicing tape on the back. Be sure, though, that no gap remains between the joined tape segments. Otherwise the adhesive from the splicing tape will come through and foul up the recording head. The marking pencil is used to mark the exact spot for cutting after you have located the spot sonically by moving the tape back and forth. (See illustrations on page 152.)

If you plan to cut and splice your tapes, you should record each tape in one direction only. The reason is obvious. If you put two separate recordings on the tape—one in each direction—you can't cut one without almost cutting the other. In short, if you flip your reels and record both coming and going, you will have to sacrifice one recording in order to edit the other. Also, it is advisable to record at the higher tape speed of 7½ ips rather than 3¾ ips, for the significant pauses between words and syllables then take up twice as much space on the tape and thus leave you more leeway for your cutting.

Multitrack Tricks

Some of the more elaborate quadraphonic recorders available today provide facilities that virtually make the recorder a self-contained sound studio. The most significant of these

Six steps for a smooth tape splice. (*Courtesy 3M Company*)

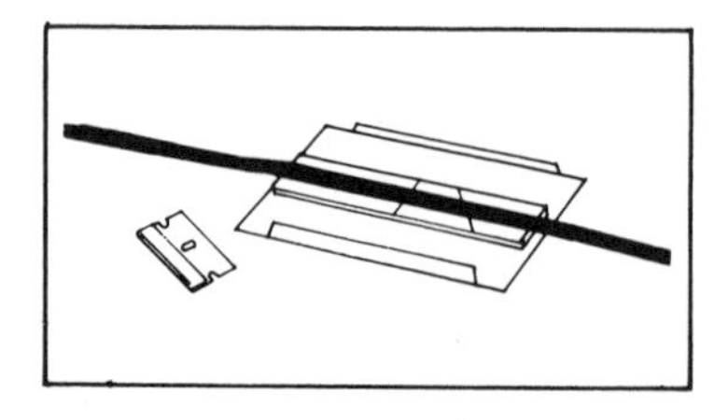

1. Place tape within splicing block guide channel, *backing side up.* Overlap ends. (Tip: Fold back end of top piece of tape and crease for easy removal after cutting.)

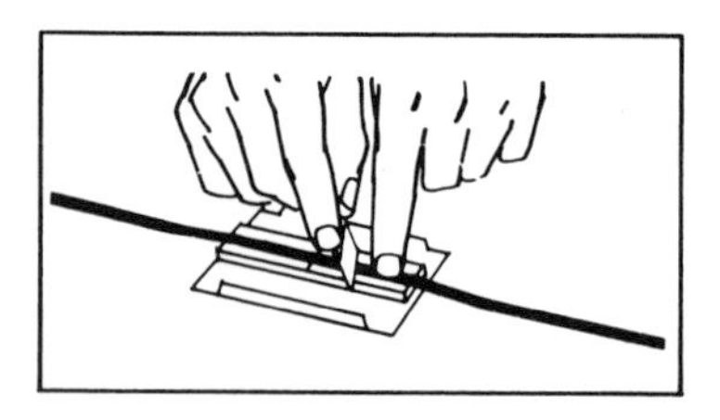

2. Hold tape firmly in channel with finger and cut with sharp, demagnetized razor blade (45° diagonal cut for regular splices, 90° cut for editing splices).

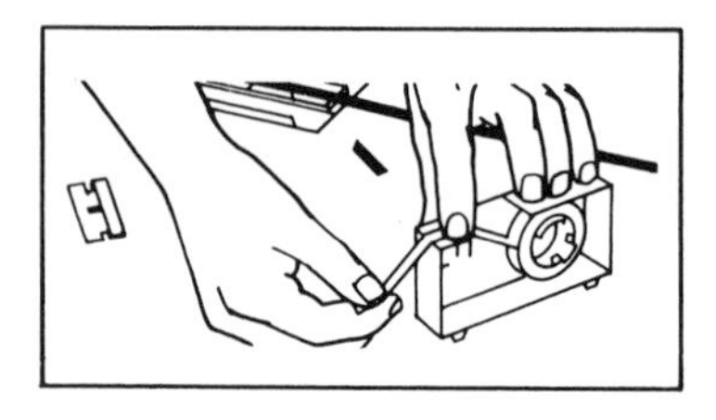

3. Pull out approximately one inch of splicing tape and draw downward against cutting blade . . . use finger to hold tape on dispenser platform to prevent slippage for cleanest cut.

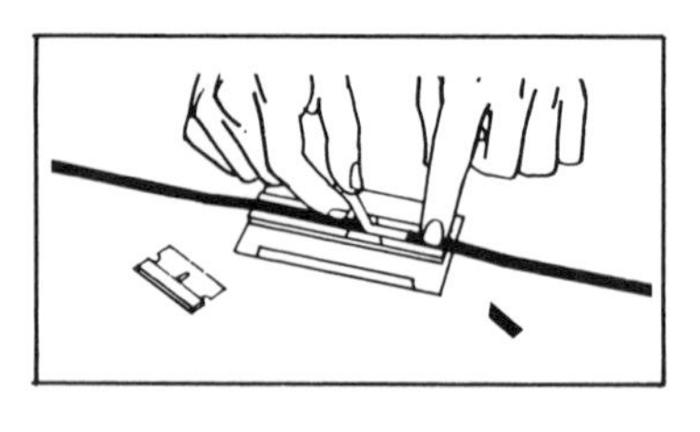

4. Keep tape ends butted tightly together and lay splicing tape carefully on top, inside channel. Press down lightly to adhere tape full length of splice.

5. Remove spliced tape from channel and rub splicing tape firmly with fingernail to remove all air bubbles.

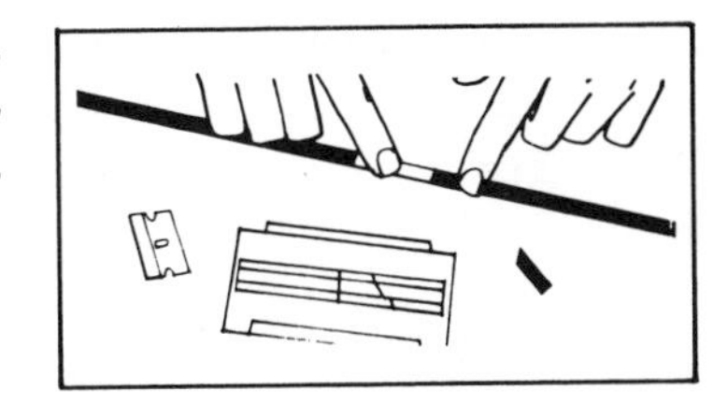

6. Clip splicing block to open side of dispenser for convenient kit storage. Keep in a clean, dry place when not in use. (If splicing tape isn't used for several weeks, remove length that has been exposed to air before next splicing use to assure fresh, clean adhesive contact.)

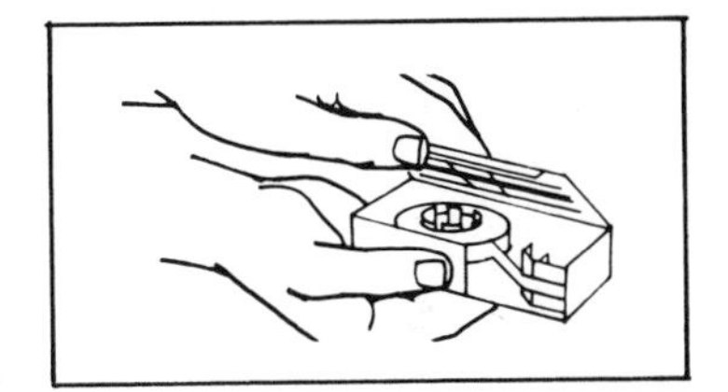

is frequently called Sel-Sync and lets you hear a previously recorded track through monitor earphones while adding new sounds in exact synchrony with what has already been recorded. Since the prices of such machines start at about $600, their advantages and challenges may well be reserved for serious hobbyists who look unflinchingly at such price tags. But if you are a professional musician—particularly in the jazz or rock area—the Sel-Sync can be very helpful if you want to make your own audition tapes rather than have them done at a commercial sound studio.

The main practical use of Sel-Sync is that it lets you add musical parts to a previous recording with accurate synchronization of all parts. That way, you can have a group of musicians play "together" without ever meeting in person. In fact, this is how many professional rock and jazz groups are recorded. The parts they sing and play are assembled from separate "takes" made at different times and perhaps in different places. The musician adding his own part to a prior recording hears the other musicians—from the earlier tape—as if he were actually sharing the stage with them. In this way, you can transform yourself into a chorus or a rock group by multiplying yourself—track upon track—on tape. Some of the recorders equipped to do this also feature a circuit to provide artificial echo, which can be added to any or all of the tracks, producing a variety of sonic effects limited only by your ingenuity.

If you are interested in such advanced recording techniques, some of the larger audio dealers will be glad to offer some instruction and let you experiment with the equipment. Some audio dealers offer special seminars for this kind of sophisticated taping.

Various companies employ different trade names for the Sel-Sync feature. Teac calls is Simul-Sync, Akai calls it Quadra-Sync, and Dokorder calls it Multi-Sync. In each case, the creative possibilities for a competent group of pop, rock, or jazz musicians are endless, and the main value of such elaborate equipment for aspiring professionals is that it enables them to experiment with their own distinctive sound patterns until they hit on what best expresses their musical imagination.

Special Sound Effects

Creative recording techniques are not the sole prerogative of those owning complex and expensive equipment. Certain sound techniques are feasible even with the simplest recorder, notably the creation of sound effects for dramatic readings of theatrical productions. Here are a few guidelines on how to produce the most common of these effects:

—Rain. Take fifteen to twenty dried peas and let them roll back and forth over a fine-meshed wire sieve while holding the microphone directly underneath the sieve.

—Wind. Pull a piece of silk across the smoothed edge of two

Some large audio dealers conduct special seminars to demonstrate sophisticated recording techniques. This meeting, sponsored by Teac in Los Angeles, shows the creative possibilities of multitrack recording. (*Courtesy Teac*)

panels of thin plywood. The gusting of the wind can be suggested by varying the speed of the pull.

—Thunder. Shake a large piece of sheet metal quite close to the microphone. If you can't find a suitable piece of sheet metal, strike the notes in the bottom octave of a piano softly and simultaneously with the palms of your hands and record them at 7½ ips. Then play them back at 3¾ ips. Naturally, this does not work with a cassette recorder or without a piano. Under those conditions, rub a piece of rough flannel or similar fabric very softly across the microphone.

—Waves. The sound of waves can be produced quite naturally by agitating the surface of water in a plastic pail (better yet, a bathtub) and recording the sound of the water lapping against the sides. Be careful not to get the microphone wet.

—Fire. To suggest a cozy fire in the fireplace, crush a matchbox about two or three inches from the microphone. For a conflagration, crush a sheet of cellophane.

—Steam Locomotive. Take two wooden blocks covered with sandpaper and rub them against one another.

—Rowboat. Dip two pieces of wood rhythmically into a pail filled with water. For added realism, let a rusty hinge squeak in unison every time you immerse your "oars."

—Pistol Shot. Strike a table with the flat side of a ruler. Hold microphone within five inches.

—Horses. Hoofbeat on pavement: take two halves of a coconut shell and strike them together.

—Jet Plane. Run an electric hair drier near the microphone and make it howl by restricting the exhaust.

—Car Crash. Bang two aluminum baking tins together.

—Gurgling Brook. Blow gently through a straw into a glass of water.

Such sound effects, with appropriate dialogue and narration, can add a new dimension to "silent" home movies even without precise synchronization, which is technically complicated. Apropos of synchronization, photo dealers sell devices for sychronizing slide projectors with a tape recorder. You may find them convenient, but I still prefer to synchronize my slide projector with my travel tapes by hand-controlling the projector so as to keep up with the taped program.

The fancy tricks discussed in this chapter cover a large variety of different applications, and I do not mean to suggest that you should get involved in all of them. The particular method or form by which you project your imagination onto tape doesn't really matter. Nor is professionalism or expertise a reasonable end in itself. What does matter is that tape—like empty canvas to a painter—offers a medium and a challenge to your creativity.

CHAPTER 12

PICKING YOUR TAPE TYPE

The most telling recent advances in recording technology have not been in recorders as such. Improved tapes have been the mainspring of progress. It is largely because of improved tapes that cassettes now rival the quality of open-reel recordings, and significant further improvements in cassette tape are announced almost every month. New formulations for open-reel tape have also been introduced recently. The sheer multitude of tape types now on the market, and the confusing and sometimes conflicting claims made for them lead to three basic questions:

1. What is the difference between one kind of tape and another?
2. Are there real tape bargains?
3. Which tape is best for what job?

As the actual carrier of the recorded sound, the tape largely determines what you hear; no recorder can be better than the tape running in it. No matter what the capabilities of the recorder, the tape itself affects such important factors as frequency response, the level of hiss, and the dynamic range.

The Anatomy of Tape

All tapes have a plastic backing—usually polyester—which is anywhere from 0.25 to 1.5 mil thick (1 mil = 0.001 inch). The thinner backing provides a thinner tape; thus a greater length (and hence a longer playing time) can be crammed within the compact confines of the cassette. On the other hand, as previously noted, the thicker tapes are stronger and more durable. The plastic backing is coated with the so-called oxide layer, which carries the sound. This consists of a binder substance in which magnetic oxide particles are dispersed. Most manufacturers guard the formula for their oxide layer as anxiously as a restaurant chef guards the recipe for his sauces, and the secret mix is responsible for the sound differences between various tape brands. In fact, tape companies deliberately tailor their tapes for the particular kind of sound they hope will appeal to their customers.

One company, for example, boasts in its advertisements that the sound from its tape contains high frequencies strong enough to shatter a wineglass. This is fine if you want to shatter glassware, and it is indeed possible to "peak" the frequency response of tape to do just that if played at extreme volume. But if you just like to listen to music (and save your wineglasses and your ears), glass-shattering propensities are not a significant measure of quality. What really matters is balanced response throughout the musical frequency spectrum, and there is considerable argument on how best to achieve this.

Iron vs. Chromium

Two types of oxide layers are currently used in cassettes. One is iron oxide (Fe_2O), which is basically ordinary rust controlled for uniformity of particle-size. The other is chromium dioxide (CrO_2). Both types have specific advantages and drawbacks. Chromium tape provides stronger highs but tends to be a little weak in the low end. The result is brilliant sound—generally preferred by rock and jazz fanciers—but the sound is often lacking in solidity and warmth. Iron oxide tape lacks these super-glossy highs, but many listeners prefer its deeper and more powerful bass. Besides, iron-oxide cassettes are cheaper.

Two basic tape formats for high-quality sound.
a. open-reel tape

b. cassette tape (the lower picture shows the opening through which the tape in the cassette contacts the recorder).

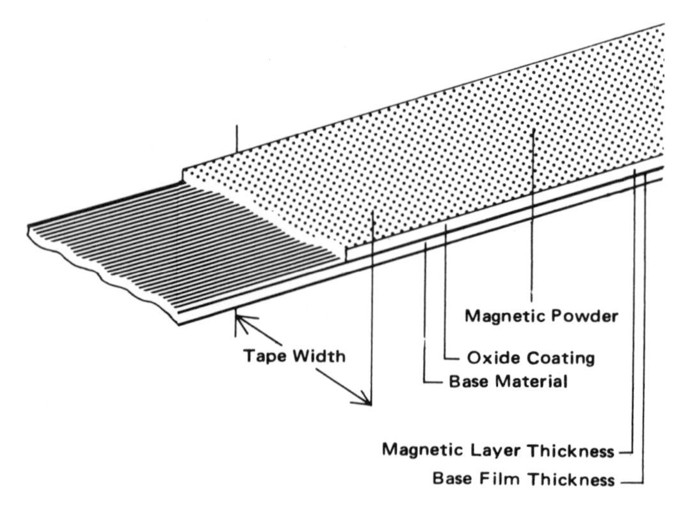

The anatomy of recording tape. (*Courtesy 3M Company*)

To get optimum results from either type requires different bias currents from the recorder. This is why the better decks feature a selector switch by which the recorder circuits can be optimized for either chromium or iron oxide cassettes. If your machine has no such switch, you can take it for granted that the machine is adjusted for iron oxide, which is regarded as the "standard" formulation.

Lately there have been attempts to combine the advantages of both iron and chromium tapes by making cassettes with a very thin chromium-oxide layer superimposed on an iron-oxide layer. This approach has been pioneered by the 3M Company in their so-called ferri-chrome cassettes. Maxell also produces ferri-chrome cassettes but on a somewhat different principle in which iron and chromium particles are intermixed in the same layer. These cassettes—based on the most sophisticated solid-state molecular technology—are priced about 30 percent higher than standard cassettes, but for those who hanker after the ultimate in cassette recording, the difference in sound may be worth the difference in price.

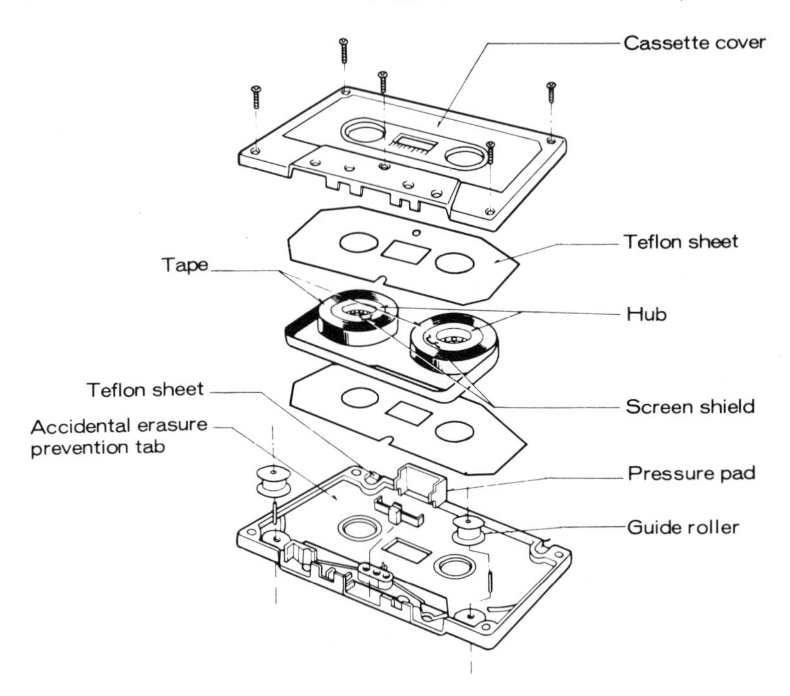

Exploded view of a cassette, showing the internal tape reels and guiding parts. In effect, the cassette becomes part of the tape transport mechanism. Accurate manufacturing tolerances and precision assembly are therefore essential to proper function. (*Courtesy Technics*)

Basic Quality Factors

Even among standard iron-oxide cassettes, there are notable quality differences. Standard cassettes contain particles of slightly larger dimensions, which are cheaper to produce. Such tapes are therefore less expensive, but their frequency and noise characteristics are somewhat inferior to premium cassettes. Still, most listeners find the ordinary standard cassettes quite satisfactory for speech recording and for the more casual kind of musical recording.

The smaller particles of the premium cassettes permit truer detail in recording, just as a fine-grained photo film allows more detail in the reproduction of an optical image. This results in better reproduction of higher frequencies. Moreover, the smaller grains can be more uniformly dispersed, which results in lower background noise, and the greater particle density permits more magnetic force to be concentrated into a given area of the tape. This, in turn, results in greater dynamic range—that is, a wider spread between the softest and loudest sound that can be effectively recorded.

Premium cassettes are also known as "low-noise" or "high-

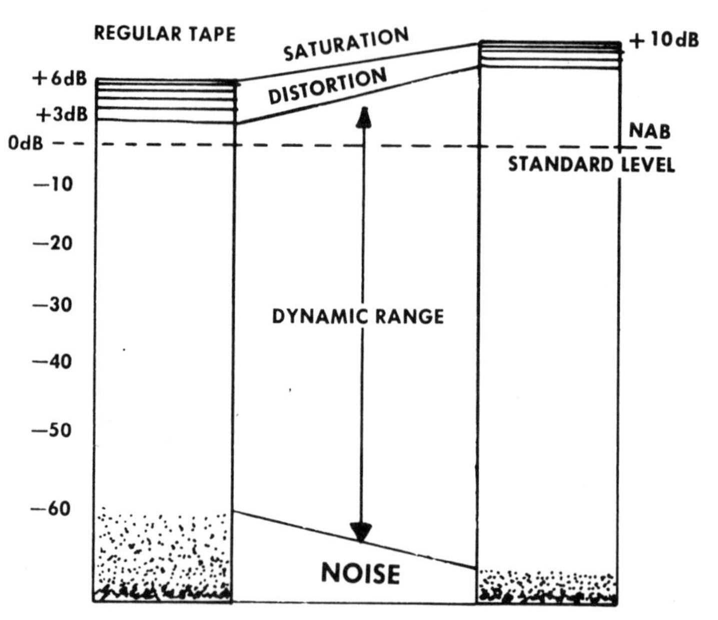

A comparison of dynamic range in standard and premium cassettes. Noise constitutes the lower limit, overload distortion the upper limit of the range. The available range on the premium cassette, charted at right, is obviously greater than that of the standard cassette, charted at the left.

energy" cassettes and are marketed under various trade names, such as the Ultra-Dynamic cassettes by Maxell, the Extra-Dynamic and Super-Dynamic cassettes by TDK, and the Classic cassettes by 3M. These companies have built their reputations on stringent quality control that assures strict uniformity of performance characteristics for every inch of the tape. Moreover, these quality brands also maintain close tolerance in the cassette shell, which helps prevent the snarling, jamming, binding, and other mechanical mishaps that often beset less carefully made products.

Bargain Cassettes

This brings us to the topic of "bargain" cassettes widely sold under numerous fanciful names at much lower prices than the standard brands. Many of them are assembled from reject batches of tape that could not pass inspection at the name-brand companies. Their magnetic coating is often uneven, which means that their sound characteristics vary from one part of the tape to another. Their frequency response may be fine in some stretches, miserable in others. Often there are so-called dropouts, where the sound vanishes entirely for a moment due to defects in the oxide layer. You may be lucky and get a bargain cassette that works perfectly. The point is, performance is unpredictable. You may indeed be getting a bargain. But you are taking a chance.

The unpredictability of bargain cassettes also extends to the plastic shell, which is an important part of the total system. When you pop a cassette into your recorder, the shell of the cassette becomes part of the tape transport mechanism. Unless the shell is machined to precise tolerance, it does not maintain uniform rolling friction. The flow of tape then becomes erratic, resulting in flutter and wow or—in severe cases—tearing, twisting, binding, and breaking. Nothing is more frustrating than to have a recording ruined by such malfunctions. I, for one, prefer paying name-brand prices simply to avoid such minor disasters and major annoyances, and regard the extra cost as a kind of insurance premium.

Another hitch with the cheapies is that an improperly constructed cassette can cause the tape to weave up and down in front of the recording and playback heads. Because cassette tape is so narrow, even the slightest deviation from strictly straight-line travel and correct alignment to the heads can make the signal fade or drop out completely at certain points. Quality cassette shells maintain tolerances as close as 0.002 inch—a necessary degree of precision for reliable performance which fly-by-night companies fail to match.

In many instances you can tell whether or not a cassette is precision-made simply by looking at the four corners. Most quality cassettes are put together with small screws, which permit precise alignment of the cassette parts. Cheaper products are usually welded together. This doesn't mean that all screw-type cassettes are equally good. But it does allow you to eliminate welded cassettes as being inherently more trouble-prone.

Cassettes come with different tape lengths. The most popular of these are the C-60 (30 minutes per side), C-90 (45 minutes per side) and C-120 (1 hour per side). The C-120 naturally contains the thinnest tape (0.25 mil) so as to accommodate the extra length, and for this reason it is more likely to develop mechanical trouble than the shorter formats. Even so, C-120s made by reputable manufacturers offer a high degree of reliability. But unless you absolutely need a full hour of uninterrupted recording time, the C-60s and C-90s will generally prove more satisfactory.

Open-Reel Tape

Because the tape tracks on open-reel tape are twice as wide as those on cassettes and the recording speed is much faster, the requirements for the tape itself are far less critical. That is why chromium dioxide, so useful in extending frequency response in low-speed cassettes, is not used in open-reel tape. All such tape uses iron oxide as its magnetic material. Yet, as in the case of cassettes, there are differences in sonic characteristics. Some tapes are designed for particularly high performance in terms of frequency range, background noise, dynamic range, and uniformity of output. On a high-quality recorder, such tapes will sound better at 3¾ ips than standard tapes do at 7½ ips, which more than offsets their extra cost. Among the manufacturers of such premium tapes, Maxell, TDK, 3M, and BASF are the most reputable brands.

No magnetic tape has an inherently flat frequency response. The highs and lows have to be especially boosted to register "on par" with the middle frequencies. This is known as equalization. Normally you need not worry about it, since it is accomplished automatically by the recorder circuits. However, some of the recently introduced high-performance tapes require a slightly different bass and treble boost than standard reel tape. Some recent open-reel recorders therefore have special switches for optimizing equalization (as well as bias) for these special tape formulations.

Among open-reel formats, you have your choice of 5-inch reels, 7-inch reels, and the professional 10½-inch reels. The length of tape wound on these reels depends on the thickness of the backing. The standard thicknesses for open-reel tape are 0.5, 1.0, and 1.5 mil. A 7-inch reel will hold 1,200 feet of 1.5-mil tape, enough to play 1 hour per side at 3¾ ips. By using 0.5-mil "triple-play" tape, you can extend the playing time per side to three hours. But the extremely thin tape is tricky to handle. To thread it into your recorder, you need the deft touch of a surgeon. Besides, your recorder has to be in top shape so it won't jerk the easily damaged thin tape when starting and stopping. The best compromise between tape thickness and playing time seems to be the "long-play" tape which has a base 1.0 mil thick and plays for 1½ hours at 3¾ ips on one side of a 7-inch reel.

Tape in the making: In this rotating drum, iron oxide particles are blended with the binder to make the finished magnetic tape coating. (*Courtesy 3M Company*)

Reel Bargains

As in the case of cassettes, the main difference between top-brand reels and bargain tape lies in quality control. The effort to make every inch of tape live up to stated specifications is what accounts for higher cost. But if you are on a tight budget, try some of the house brands offered by some of the larger electronic mail-order companies which also maintain numerous retail outlets, such as Radio Shack or Lafayette.

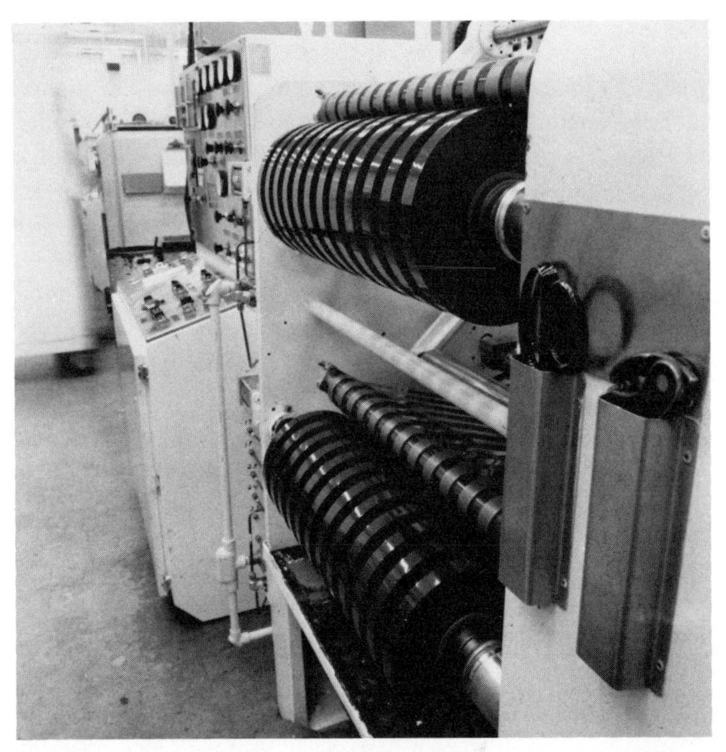

The tape is slit to exact width on these precision machines. Maintaining constant tape width over the entire length of the reel is one of the keys to satisfactory performance. (*Courtesy 3M Company*)

Chances are that you'll get very good value for your dollar and far better reliability than on strictly off-brand tapes. The same holds true for cassettes offered under the house-brand names of such firms.

Yet there are some bargains that are no bargain at all. These are called "white-box" tapes because they often come in white, unmarked boxes or are sold under spurious brand names designed to hide rather than identify their source. You may be able to buy five such reels for the price of a single standard-brand reel. If you're lucky, you are getting reject computer tape in those white boxes. It's designed for very high frequencies—about two hundred times higher than the highest audible sounds—to record the rapid pulse sequence of computers. Consequently, its response in the audio spectrum is very uneven and particularly weak in the bass. Even so, speech will sound intelligible, though the timbre of the voice will be falsified.

But occasionally the white box contains reject TV tape. That's another story. Unlike audio tape, video tape has all the magnetic particles standing upright instead of lying down sideways. The tape may carry the sound all right, but it will hiss at you like a snake.

No tape is a bargain if it damages your recorder. And some off-brand tapes may do just that. Unless a tape is impregnated with a silicone-type lubricant, it may grind down the metal surfaces of your magnetic heads. Luckily, poorly lubricated tape usually sounds a warning as it runs through your recorder: it squeals.

Other tapes may be badly lubricated so that their surface is sticky, causing adhesion between the layers on the reel. This keeps it from unwinding freely, causing flutter and wow because tape tension varies as the tape pulls loose from the sticky spots. One way to check for adhesion is to take the reel and simply unwind it in your hand until about twenty to thirty inches hang down toward the floor. If this length spins off without sticking, chances are that the rest is okay, too.

Poor tape also sheds its oxide particles as it rubs against the recording and playback heads. The metal dust makes a fine grinding compound that soon causes the finely machined edges of the head-gaps to become ragged. Besides, the oxide dust clogs the heads—and a clogged head does for tape recorders what a stuffy nose does for people: it makes them sound pretty awful. To test whether your tape sheds its metal grains, run a fingernail over the coated side of the tape.

If any brown powder collects under your nail, the tape is suspect.

I have gone into gruesome detail on the various failings of tape to urge upon you an attitude of *caveat emptor* when it comes to tape and cassette "bargains." After all, there is no point in handicapping a fine recorder with poor tape, and only a well-engineered, reliable tape can assure the satisfactions you rightly expect from good equipment.

CHAPTER 13

MAINTENANCE AND ACCESSORIES

Like any mechanical device, a tape recorder needs occasional maintenance to keep it in top condition, and a few simple routines will help to assure proper performance for years.

Cleaning

Dust getting into the moving parts of a recorder is more ruinous than normal wear. Moreover, dust grinds the highly polished metal surfaces that come into contact with the moving tape. This abrasion interferes with the even flow of tape, causing an increase in flutter. Moreover, as dust grinds down the recording and playback heads, frequency response suffers.

To forestall such deterioration, clean all metal parts contacting the tape by wiping them gently with a cotton-tipped Q-tip. If you use your recorder daily, this should be done about once a week. In cleaning the heads, draw the Q-tip vertically over the head in the direction of the gap—never draw it across the gap, as this will only force the dust deeper into it. Also, never touch the metal surfaces of the heads with your fingers, for the natural skin oils contain corrosive substances that will permanently damage the highly polished surfaces. The other metal parts that guide the tape through the mechanism as well as the capstan and the rubber idler that pull the tape past the heads should also be cleaned carefully with cotton.

Some manufacturers recommend cleaning these metal parts with isopropyl alcohol to remove oily film. But this is a somewhat risky prcedure, for any alcohol running down into the bearings would dilute the lubricating fluid. If you use alcohol, be careful that none of it touches the rubber idler that holds the tape against the capstan, for the alcohol may harden the rubber and cause it to slip over the tape rather than pulling it evenly. Unless you are fairly deft, you had better stay away from alcohol and give your recorder a dry cleaning.

You can gain better access to the heads for cleaning purposes if you put your recorder in the PLAY mode (while keeping the power switch off), because this shifts all the heads forward and makes them easier to reach.

Regular cleaning of this sort also removes any stray oxide particles that accumulate in the head gaps and cause noticeable loss of high frequencies.

Demagnetizing

In the course of normal operation, as magnetized tape flows through the recorder during playback, tiny traces of magnetism are imparted to the heads and the other metal parts of the transport mechanism. These tiny magnetic charges gradually build up and cause a noticeable increase in hiss. The unwanted magnetism also tends to wipe off the high frequencies on previously recorded tapes, so that your tapes can be permanently damaged in playing unless you demagnetize your recorder about once a month.

For this you need a special device known as a demagnetizer or "degausser," the latter name commemorating Friedrich Gauss, the German mathematician and physicist who explored magnetic phenomena in the last century. Such devices look like an electric soldering iron and can be obtained for a few dollars. They are virtually indispensable if you want to keep your recorder and your tapes in prime condition, but they must be handled with a certain amount of caution. Otherwise they do more harm than good and may leave your recorder even more magnetized than before.

The main thing to remember is never to turn the demagnetizer either on or off when it is near the recorder. The reason for this is that the moment of switching generates an extra-strong magnetic surge that permanently magnetizes anything nearby. The correct procedure is as follows and takes less than two minutes:

1. Turn the recorder off, remove the head covers so as to expose the heads to full view, and push the PLAY button to bring the heads forward.

2. Hold the demagnetizer at arm's length away from the recorder (and any tapes that may be lying around), turn it on and then bring its tip slowly toward each of the tape heads and other metal parts of the tape transport, moving it slowly up and down over the heads and along the area through which the tape passes. It is not necessary to touch the head surface directly, and you should avoid doing so. The demagnetizer works effectively if held about half an inch from the head surface. The tip of the demagnetizer should be covered with cellophane tape or some plastic to avoid the danger of scratching the heads if you should accidentally touch them.

3. After several passes over the heads and metal parts, slowly withdraw the demagnetizer and don't turn it off until it is again several feet distant from the recorder.

This simple procedure will leave your machine in magnetically pristine condition, assure hiss-free, wide-range recordings over a long period of time, and protect your tapes from being partially erased while playing.

If you own a cassette recorder, the procedure is basically the same. Yet there is an even simpler alternative. You can buy a special cleaning cassette which snaps in just like a regular cassette. But instead of producing sound, it cleans the heads as you "play" it. Some of these cassettes also have magnetic particles imbedded in them in such a fashion as to neutralize magnetic charges that have built up in the head. In short, these cassettes both clean and demagnetize your recorder. Still, I suspect that a regular head demagnetizer used as described above will do a more thorough job, as will cleaning with a Q-tip.

Major Maintenance

Most high-quality recorders are built to maintain their mechanical parameters constant through many years of use. Their frames are precision castings that keep their dimensions under normal use. But if the machine is subjected to abnormal stresses, if it is handled roughly or accidentally dropped, these basic mechanical adjustments may be disturbed. Mechanical shocks may affect the critical head alignment, which keeps the head gaps perfectly perpendicular to the direction of tape travel and keeps the head at precisely the proper height in respect to the tape. If the alignment has been thrown off by some shock or accident, it is possible to realign them with the aid of so-called alignment tapes. But to do this properly requires test instruments, and it is not a procedure to be entrusted to amateurs. It should be carried out by a factory-authorized service agency. The same holds true for adjusting the tension of the drive belts that operate the reels and the tension springs pushing the tape against the heads.

Accessories

Your "housekeeping" associated with tape recording will be greatly simplified with some readily available accessories, such as proper storage boxes for tapes and cassettes. These not only make it simpler to organize your tape library but also help keep your tapes and cassettes protected from dust.

We have already mentioned such other accessories as microphone mixers, earphones, splicing blocks, splicing tape, and leader tape in connection with their respective uses. But earphones require some additional discussion.

Earphones

So far, earphones have been mentioned only in connection with on-location recording, where you need them to monitor your "take." But, of course, they are also immensely useful at home, where they provide opportunity for private listening without disturbing other members of your household. With earphones, privacy works both ways; not only does it shield the members of your family when you want to listen to music at full volume while they, possibly, may have other things on their minds. It also allows you full concentration as you listen.

Earphones have the peculiar effect of making you virtually oblivious to your surroundings. Psychologically speaking, they remove you from the scene. Slip them on and you suddenly feel whisked away from all the petty distractions that impinge on you in your home, and you are left alone with the music. Because the sound goes directly to your ears, it skips all the acoustic quirks of your living room. Problems like speaker placement and stereo listening location are automatically bypassed. The acoustic ambience of the place where the original recording was made reaches you without being altered by your own home acoustics. The net effect is an uncanny illusion: the very space of the concert hall or recording studio seems infused via the stereo earphones right into your head.

Donning stereo earphones, you can "feel" the whole concert stage—maybe some sixty feet wide—spreading out in whatever space there happens to be between your two ears! With the whole orchestra inside your cranium, you'll find it hard to believe that your hat size stayed the same.

How does this happen? Nobody knows for sure. The basic process of sensory perception is still not completely understood. Psychologists, physicists and philosophers have yet to discover the exact relations between objective reality and our senses. Of course, it may be argued, as did Bishop Berkeley and Immanuel Kant, that all perception ultimately lies within our heads. The sensation of hearing stereo "space" via earphones is surely an astounding demonstration of this.

Getting back to the more tangible subject of hardware, earphones tend to have their individual sound coloration, which varies from one model to another. In selecting a model for your own use, compare different designs just as if you were buying a pair of loudspeakers. Clarity is the most important criterion. Make sure the sound doesn't blur even at full volume. Try to pick out the individual instruments of the orchestra. Watch for the presence of bass even in soft passages; note whether the sound of the lower strings—such as cellos and basses—has the proper solidity. Check the crispness of sound in such instruments as harpsichord, guitar, and various kinds of percussion. And watch for the common drawback of inferior designs—high-frequency distortion. Violins, for example, should sound silky and smooth, without stridency.

Aside from sound, fit is the main factor in picking your earphones. You should be able to wear your headset all evening without any discomfort. Fit around the head is rarely a problem because most headbands are either flexible or otherwise adjustable. The earpieces, however, have fixed dimensions. So make sure they don't pinch or squeeze your ears. They should fit around the ears rather than over them. They should also provide a good air seal; otherwise you lose bass response. Some ear-cushions are liquid-filled so that they mold themselves to the contours of your head. Others rely on foam materials to form an efficient sound seal. On some models, ear cushions are washable, a decided advantage in case they become grimy or saturated with skin oils.

The lighter the headset, the longer you can wear it without fatigue. For this reason, all recent models are made of lightweight materials. Some weigh less than 10 ounces. You are hardly conscious of wearing them.

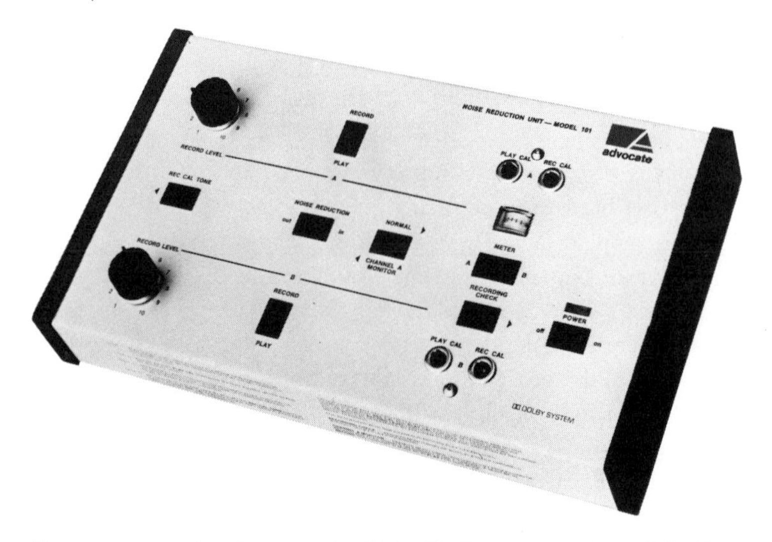

If your recorder has no built-in Dolby, an external Dolby device such as the one shown here can be connected to it for greater noise reduction. (*Courtesy Advent Corp.*)

If you are among the avant-garde owning quadraphonic equipment, you can also obtain four-channel earphones which contain two separate sound elements in each earcup so as to simulate the quadraphonic effect by providing two separate signals for each ear. One represents the front and the other the rear channel.

Bulk Erasers

There is still another accessory that may prove useful if you have many tapes and cassettes that you want to record over and over again. This is the bulk eraser, which lets you erase a complete reel or cassette within seconds, without having to run it all the way through the recorder. The bulk eraser has yet another advantage if you use the same tape repeatedly for different recordings. A frequently rerecorded tape or cassette retains traces of residual noise that the ordinary automatic erasing process of the recorder does not quite eliminate. Bulk erasers, by contrast, apply a much more powerful demagnetizing force than the erase head of the recorder, capable of wiping a tape perfectly clean of all prior sound traces. Thus, with the aid of a bulk eraser you can make even your old tapes which have been used many times over as silent and receptive for new recordings as when they first came from the factory.

A bulk eraser, costing $20 or thereabouts, is a strong electromagnet which is passed swiftly across each side of the reel or cassette. Yet as in the case of head demagnetizers, it must be used with caution. You should not switch it on or off in the close vicinity of the tape or cassette to be erased because the strong switching pulses leave a permanent magnetic imprint. Rather, switch on the eraser while holding it at some distance from the tape. After passing it over both sides of the tape, remove it very slowly while it is still on and then switch it off at a safe distance again.

These simple maintenance and "household" routines may seem somewhat trivial at first glance. Yet they are of inestimable importance in keeping your equipment functioning at its best and assuring consistently good results.

CHAPTER 14

BETTER SOUND FOR PLAYBACK

The main concern of this book is with recording equipment and techniques. To deal with the components needed for playback—amplifiers and loudspeakers—would require me to write a separate book. Actually, I have already done so. If you are interested in acquiring a basic sound system to make the most of your tapes and cassettes, I would suggest that you read my book entitled *The* True *Sound of Music,* also published by E. P. Dutton and Company. Any bookshop will order it for you on request.

For the present, let us assume that you already own a good stereo or quadraphonic sound system on which to play your tapes. The following suggestions (adapted from *The* True *Sound of Music*) will help you make the most of the equipment you already have.

Speaker Placement

One of the most important factors in playback is speaker placement. In fact, one of the great assets of components—compared to all-in-one consoles—is that they permit you to locate the speakers at that place in the room where they sound best. It is surprising how many components owners don't take full advantage of this fact. Since optimum placement in a given situation depends on the size, shape, and furnishings of your listening room and the sonic characteristics of your speakers, there can be no hard and fast rules to fit all cases. The major consideration, of course, is getting the most natural and balanced sound that your speakers can achieve.

You can get more bass, for example, just by moving the speakers into corners. Putting the speakers right down on the floor (the woofer end of the cabinet [the end with the bass speaker] should be in the lower position) also reinforces bass. The reason is that the wall and floor surfaces adjoining in the corner help project the lower frequencies more efficiently into the room, almost like a horn. You can achieve the same bass-boosting effect by mounting your speakers in ceiling corners, a handy arrangement in small rooms with a shortage of floor space. Wall brackets serve nicely for this purpose.

Bass-reinforcing placement may be a boon to inexpensive systems whose smaller woofers tend to be bass-shy. However, with full-size speakers, such placement may result in an overly heavy, unbalanced bass. In that case, it is best to keep the speakers out of corners and perhaps off the floor.

To operate efficiently in the lower bass range, a loudspeaker should be at least fifteen to seventeen feet from the wall toward which it faces. This allows a sound-projection path long enough to accommodate half the wave lengths of the lowest musical notes and lets these deep tones come through more powerfully. That is why, especially in rooms of moderate size, you often get richer sound by placing the speakers against the short walls so that they face the full length of the room.

Since the better speaker systems available today have good high-frequency dispersion—they radiate the treble frequencies over a broad angle—placement is not super-critical as far as the higher treble frequencies are concerned. However, if your speaker system seems to lose highs as soon as you sit down, you had best set up the speakers so that the tweeter (the high frequency speakers) are at ear level.

Room Acoustics

Room acoustics is a second factor to consider when placing speakers. In addition to their effects on the overall balance of highs and lows, these resonances color to a greater or lesser degree all the sound produced by your speakers. Every room has resonant peaks and dips at certain frequencies and locations, depending on the room's dimensions, giving it what might be called tonal personality. The tonal character of a room is determined by its size, its shape, and its sound-reflective qualities. These factors decide the duration of an echo in the room—acoustic engineers call this reverberation time—and the pitch of the notes that are predominant in the echo. A long reverberation time makes for a spacious sound, and the more reflective surfaces there are, the more the higher tones are emphasized.

Hard surfaces, such as plaster walls, tile floors and windows, reflect sound the way a mirror reflects light and make a room acoustically "live." The proper amount of reflection gives the sound a pleasing brilliance and richness. Too much reflection causes shrillness and jumbles the music unpleasantly by prolonging each note with excessive echo. Moreover, the music seems to come from all sides at once. The other extreme, too little reflection, makes the sound lackluster and dead.

As a rule, you cannot alter the size and shape of your listening room and thus cannot control its reverberation time. Fortunately, the reverberation effect of the concert hall is—or should be—contained on your records, so that the impression of spaciousness can be obtained even when you play records in rooms of moderate size. What you can do, though, is experiment with the balance of sound reflection and absorption in your room. If the music seems overbright and has a harsh, ringing quality, you need something to soak up some of the sound. Put a hanging or a wall rug over the wall that faces the speakers, or put up some heavy curtains. Anything that is soft will help: pillows, overstuffed furniture, rugs, and the like. These cut down the amount of sonic energy bouncing about the room and suppress excess high-frequency tones.

If the music seems stifled, lacking tinkle in the highs, try pulling back a rug to expose more floor area. Or you can take down some draperies or put up a large, glass-covered picture. Such simple measures can accomplish remarkable results.

In some rooms, problems of resonance make certain notes (usually in the bass range) sound louder than the rest of the music. This is caused by so-called standing waves that make the room act as a resonator. Sometimes it is possible to prevent the formation of standing waves by angling the speakers so that they do not project sound parallel with or at right angles to the walls. Try also to place your speakers where they will be least likely to provoke undesired resonance, or arrange your preferred listening chair so that it is not in a resonant area. This may mean shifting the speakers and the chair along the long walls of a room (or moving them away from the walls) until you find the location at which there are the fewest problems.

The last factor to consider is stereo separation. The speakers should be at least eight feet apart. You'll get the best stereo spread if your listening chair is about equidistant from both speakers. But you can still get a fine stereo effect even if you're sitting off-center. In fact, with good speakers, you'll notice the stereo spaciousness almost anywhere in the room, and therefore you don't have to be hidebound about this kind of musical geometry. Besides, the amplifier's balance control will permit adjustment of the relative speaker outputs if the layout of your room makes radically off-center listening more convenient.

Listening Location

In searching for the optimum stereo effect, the general rule is to separate the two speakers so that they subtend an angle of roughly thirty to forty degrees as seen from the listening position. But since each room has its individual characteristics and since furniture arrangements are rarely alike, this general rule is subject to all sorts of variations. More often than not it serves merely as a starting point for experiments. For instance, you can put the speakers farther apart and compensate for the added separation by means of the blend control provided on some amplifiers. Or, if your room is so narrow that you cannot separate the speakers far enough, you can place them at right angles against two adjoining walls with the listening area approximately at the intersection of the two sound-projection lines.

It is also possible by means of speaker placement to emphasize either the directionality or the depth of sound that together make up the stereo effect. Directionality is stressed if the speakers face directly toward the listener. But if your preference runs toward greater depth of sound, with music seeming to fill the whole room without a discernible source, try omnidirectional speakers, or angle your conventional speakers outward toward the nearest wall so that they face away from each other and their sound reaches the listener only on the rebound. This method is especially effective in enhancing the sense of acoustic spaciousness in small rooms, although some stereo separation may be lost. In high-ceilinged rooms or under a gabled roof you can even turn your speakers on their backs so that they face upward and their sound is reflected from above. Though this also reduces stereo separation, the use of reflected sound widens the area of the stereo effect so that the location of the listener becomes less critical.

Not that the listener position for stereo is as critical as many people think. In stereo's early days, it was a commonly accepted half-truth that the listener had to sit at equal distance from the two speakers to hear the maximum stereo effect. Such a listening position is comparable to a center-aisle seat in the concert hall, and it puts you in the location with the most balanced right-and-left sound distribution. In a room of average size, the stereo speakers might be placed from eight to twelve feet apart, and you could then put your favorite chair across the room from the loudspeakers at a point somewhere close to an imaginary line drawn midway between them.

But you needn't be dogmatic about your stereo seating arrangement. Stereo permits the listener far greater freedom of movement than orthodox stereophiles will admit. In fact, the stereo effect can be appreciated almost anywhere in the room. If you are sitting off-center, your location might be compared with a box seat along the side of an auditorium, which is nothing to complain about. You may perceive less left-right directionality in such a location, but the essential fullness and spaciousness of stereo will still be retained.

To prove to yourself that the stereo effect is not strictly localized, just walk across the room in front of the speakers while a record is playing. It's like dancing across a ballroom in front of the bandstand: although the sound perspective changes at various points, depending on which instruments you are closest to, these are quite natural changes and the stereo effect is not lost.

Spreading stereo sound evenly over a wide area requires a wide angle of treble dispersion from your speakers. Loudspeakers differ considerably in their ability to fan out the high frequencies. The poorer ones project the treble in a narrow beam, as from a flashlight. This leaves at the sides of the beam large areas of aural "shadow" in which the sound is dull. In general, broad-angle dispersion of highs provides greater latitude in the choice of listening location.

Four-Channel Placement

Although the discussion of speaker placement applies just as much to four-channel as to stereo installations, quadraphony does add some factors of its own. The main problem is arranging your listening position so that the rear speakers are balanced in respect to the front speakers. Moving your listening chairs to the center of the room would be the ideal solution. If your living room is big enough, I recommend it. But for most of us, with average-sized rooms, a clump of furniture in the room's center would entail insurmountable traffic problems, not to mention aesthetic atrocities.

Practical room arrangements usually seat at least some listeners along one wall, and we must allow for this. I've gotten good results by placing the rear speakers right along the couch on which I listen, with the front speakers facing me across the room. But instead of facing the rear-channel speakers straight forward, I angle them outward toward the side walls (away from the couch). Then I adjust the volume so that the sound from the rear speakers does not overbalance the front speakers.

Since every living room presents an individual acoustic situation, you may have to experiment a little to discover the speaker placement that is best for your particular circumstances. Don't try to solve this problem in a hurry. Try different speaker positions and live with each of them for a few days. Eventually you will find the one that most closely approaches the ultimate aim of good audio: to let you forget all about the equipment, to make you unaware of its presence, so that you feel there is nothing between you and the music.

APPENDIX

THE NATURE OF SOUND

Throughout this book we have been concerned with the nature of sound. But we have discussed it mainly in its practical aspects as related to specific recording tasks. If you are eager to gain a somewhat more theoretical understanding of the phenomena of sound and hearing, the following presents a brief outline of basic concepts that relate sense perception to physical phenomena.

It is all too easy, in discussing the principles of sound reproduction, to take for granted—and therefore lose sight of—the first requirement for high fidelity: something to be faithful *to*. Sound, after all, is not only the end product of high fidelity, but the raw material as well. The central purpose of an audio system can perhaps best be described as the accurate reproduction of that part of the infinitely various world of sound that is used for the creation of music.

Take, for example, the matter of tone color—also called *timbre*—which contributes so significantly to the expressiveness of music. A cello and a trombone may both be playing the same musical note, but the listener can easily tell them apart. What accounts for this?

When the great German physicist H.L.F. Helmholtz first began to analyze sound some hundred years ago, he discovered that what the listener hears from an instrument as a single musical note actually consists of many different tones. There is, first of all, the basic pitch perceived by the ear—called the *fundamental*. But in addition to this fundamental tone, the musical note embodies a whole series of additional tones, called *overtones* or *harmonics*. These are multiples of the fundamental frequency (i.e., twice, three times, or four or more times the frequency of the basic note). Not all of these overtones are equally strong, nor do they have the same phase (time) relationship to the fundamental. Each musical instrument has its own individual overtone pattern and it is this pattern that gives each instrument the characteristic tone color by which we identify it.

Now, since audio components and tape must reproduce this pattern accurately if the sound is to be faithful to the original, it can be seen that a wide—and uniform—range of frequency response is necessary. Suppose an oboe is playing a note with a basic frequency of 1,500 Hz. Its overtones would be 3,000 Hz (1,500 × 2), 4,500 Hz (1,500 × 3), 6,000 Hz (1,500 × 4), and so on. To do justice to the sound of the oboe, the system must reproduce these overtones in the exact strengths and phase relationships in which they occurr in the original.

Even instruments such as the bass viol, the tuba, and the kettledrum—the lowest-pitched instruments in the orchestra—produce higher frequency overtones that give them their particular tonal flavors. This accounts for the seeming paradox that a sound system must have a frequency range up to at least 15,000 Hz in order to reproduce instruments whose basic pitch is in the lowest octaves of the musical range.

Another requirement for faithful reproduction of timbre is the absence of harmonic distortion. This type of distortion occurs when an electronic component, in effect, adds to the sound some overtones that were not contained in the original music. Such additions falsify the tone color because they alter the overtone pattern of the original. Modern audio circuits are carefully designed to keep this type of distortion at levels so low that the effect is seldom perceptible at normal playing volumes.

Any discussion of the principles of sound must also include a brief explanation of its two fundamental attributes—pitch and loudness. These, together with harmonic structure, are the elementary physical realities of musical sound, the absolutes against which the performance of an audio system must be judged.

From one point of view, sound may be considered a purely subjective phenomenon, a sensation experienced principally by the ear and, to a certain degree, the entire body. But to understand the technical aspects of audio, we need to relate this subjective experience to the corresponding objective physical events that produce it. Sounds are fast-moving areas of high and low air pressure created by the "pumping" action of a sound source such as a plucked string, the vibrating body of an instrument, or the resonant action of an enclosed air mass. The air adjacent to these vibrating elements is driven back and forth, thus forming a series of pressure peaks and nulls that are first sensed by the eardrum and are then passed on and perceived as sound by the brain.

Whenever you hear a musical note, it has a pitch—high, medium, or low. The sense of pitch is the brain's way of recognizing, without actually counting, the number of air-pressure pulses impinging on the ears within a given period of time. This number is called the *frequency* of a tone, and each vibration (or pulse) is called a cycle. Frequency—which we perceive as pitch—is therefore expressed in cycles per second, abbreviated Hz. The faster the vibrations (the higher the frequency), the higher in pitch the tone will appear to be.

The lowest sounds audible to the human ear (the roll of distant thunder, for instance) have a frequency of about 16 Hz. The lowest musical notes, such as the pedal notes of an organ, are in the vicinity of 30 Hz, although it is a rare musical note that falls below 50 Hz. A sound system with a bass response extending down to the 40-cycle range is therefore quite adequate for music reproduction.

The upper limit of human hearing varies with age. As a rule, only the young can hear frequencies above 20,000 Hz. The limit for adults usually lies around 16,000 Hz, declining with age (in most cases) to 10,000 cycles or less. No musical instrument has a basic pitch higher than about 5,000 Hz. But music played on a radio or phonograph that is limited to a top frequency of 5,000 Hz sounds stripped of tonal richness—and it is, for the overtones that lend each instrument its characteristic tone color fall mainly in the 8,000- to 15,000-Hz frequency range. A frequency response that extends to at least 15,000 Hz is therefore essential if the reproduced sound is to bear any relation at all to the original. Audio engineers are still disputing how high an amplifier's frequency response must go in order to reproduce accurately the waveshapes of transients and to preserve the proper time relationships among the harmonics.

Certain sounds do not have a definite pitch—for example, splashing water, rushing air, or wooden rattles. The reason for this is that their vibrations do not have a regular rate of recurrence, but are made up of a random mixture of frequencies that the ear is not equipped to unscramble and perceive as definite and separate pitches.

In addition to pitch and timbre, each musical note also has a certain loudness. And this, of course, means that there is yet another basic requirement for audio equipment: it must be able to render different loudness levels in natural proportions, doing equal justice to delicate *pianissimos* and to riotous orchestral outbursts. The dynamic range of broadcast or recorded music (i.e., the spread of loudness between softest and loudest passages) should have no effect on the quality of the reproduction. Loud passages should not drive the amplifier into distortion and soft passages should not be accompanied by amplifier hum, turntable rumble, or similar nonmusical distractions.

Thus, to reproduce loud notes adequately and with low distortion, particularly the deep bass notes of the low-pitched instruments, the amplifier needs sufficient power, i.e., wattage: the greater the electrical energy, the louder the sound. Power requirements and other aspects of playback equipment can be mentioned only briefly in this context. A more detailed discussion of playback equipment can be found in my book *The* True *Sound of Music,* which may be read as a companion volume to the present book.

All the main attributes of sound we have described—pitch, timbre, and loudness—can be represented as wave patterns. Sound waves may be thought of as being roughly similar to waves in water. Each wave plus its accompanying trough represents a single cycle. The time interval between successive cycles (i.e., the speed of the wave crests past a given point) determines the frequency, or pitch; the shape of the wave determines tone color; and the height of the wave, which corresponds to its energy content, determines loudness. In the terminology of physics, the height of the wave is called its amplitude, and the greater the amplitude, the louder the sound.

To many people the notion of music running through a wire seems rather uncanny. The change of sound to electricity and vice versa is indeed a marvelous process, even though it is a commonplace event that happens every time you pick up the phone. The relation of sound to a corresponding electrical signal is, in fact, the key to the mysteries of audio.

At the risk of being somewhat inaccurate in the particulars, let me suggest a mental picture that you may find helpful in visualizing just what goes on in these transformations from sound to electricity and back again. Think of the electric current as a conveyor belt moving at high speed. If you can make this "belt" undulate exactly like the sound waves you want to transmit, it then becomes a carrier of sound. Thus transformed into an electric signal, music and speech can be amplified, recorded, or sent out via radio.

A GLOSSARY OF TAPE-RECORDING TERMS

This glossary of technical terms employed in tape recording has been prepared by the 3M Company, one of the major manufacturers of recording tape, and I am grateful for their permission to include it in this book. Not all of these terms are used in the text of this book, but the reader may encounter them elsewhere.

Automatic Reverse—The ability of some four-track stereo tape recorders to play the second pair of stereo tracks automatically (in the reverse direction) without the necessity for interchanging the empty and full reels after the first pair of stereo tracks is played.

Automatic Shut-Off—A device (usually a mechanical switch) incorporated into most tape recorders that automatically stops the machine when the tape runs out or breaks.

Azimuth Adjustment—The mechanical adjustment of a magnetic head whereby exact alignment of the head gap with a standard tape-recorder magnetic pattern is achieved. Of prime importance for optimum high-frequency performance and recorder-to-recorder playback compatibility.

Backing or Base—The flexible material, usually cellulose acetate or polyester, on which is deposited the magnetic-oxide coat that "records" the taped signal.

Bias—A constant signal or tone added to the audio signal during recording to circumvent the inherent non-linearity of magnetic systems. The best (and most commonly used) bias is a high-frequency (usually 50,000 to 100,000 Hz) alternating current fed to the recording head along with the audio signal to be recorded.

Bulk Eraser or Degausser—A device used to erase magnetic tape without removing it from the reel. It generally produces a strong alternating magnetic field that neutralizes all previously recorded magnetic patterns on the tape.

Capstan—The driven spindle or shaft in a tape recorder—sometimes the motor shaft itself—which rotates against the tape (which is backed up by a rubber pressure or pinch roller), pulling it through the machine at constant speed during recording and playback modes of operation. The rotational speed and circumference of the capstan determine tape speed.

Cartridge—A sealed plastic container that holds tape. Designed to eliminate manual tape threading, cartridges operate on the continuous-loop (single hub) principle.

Cassette—A device containing tape and reels that can be snapped into a recorder for convenient recording and playback without tape threading.

Decibel—Abbreviated "dB" or "db," it is a relative measure of sound intensity or "volume." It expresses the ratio of one sound intensity to another. One db is about the smallest *change* in sound volume that the human ear can detect.

Deck, Tape—A tape recorder designed specifically for use in a high-fidelity music system. It usually consists only of the tape-transport mechanism and preamplifiers for recording and playback. It does not include power amplifiers or speakers.

Dolby—An electronic device or circuit that reduces the amount of noise (principally tape hiss) introduced during the recording process. It does this by boosting—in carefully controlled amounts—the strength of weak signals before they are recorded. During playback the signals (and the noise) are cut back by an exactly equivalent amount. The original dynamics are thus restored, but the noise is reduced by 10 to 15 db. At one time found only in recording studios, simplified Dolby circuits designed especially for tape recording are now available to the audiophile as accessories or built into tape machines.

Dropout—During playback, the momentary loss of a recorded signal resulting from imperfections in the tape. These may take the form of non-magnetic foreign particles imbedded in and flush with the tape's surface. However, these imperfections are most commonly high spots on the tape surface that push the tape away from the magnetic head, thereby increasing the area affected (the "umbrella" effect).

Dub—A copy of another recording.

Dynamic Microphone—An electromagnetic pressure microphone that employs a moving coil in a magnetic field to convert sound pressure to electrical energy in a manner similar to that of an electric generator. Impedance and output are generally lower than those of the ceramic or crystal microphone types. Low impedance permits the use of longer connecting cables without high-frequency loss or hum pickup.

Dynamic Range—The voltage ratio (expressed in decibels) between the softest and loudest sounds a tape recorder or other device can reproduce without undesirable distortion in loud passages and excessive noise in soft ones.

Echo—A special facility found in some three-head tape recorders. Part of the slightly delayed output of the monitor head is fed to the recording head and mixed with the signal being recorded. The result is an "echo" of the material recorded a moment before.

Editing—The alteration of a tape recording by physical means to eliminate or replace undesirable portions, add portions not present in the original, or otherwise rearrange the original. Magnetic tape is unsurpassed for editing purposes, since it can be easily cut and spliced.

Equalization—The selective amplification or attenuation of certain frequencies. Also refers to recognized industry standards for recording and reproducing "characteristics" (such as the NAB Standard), the proper use of which can assure uniform reproduction of prerecorded tapes and improvement of a system's signal-to-noise ratio.

Erasure—The neutralization of the magnetic pattern on tape by use of a strong magnetic field, thereby removing the recorded sound from the tape. During recording, the erase head on a recorder automatically removes any sound previously recorded on the tape just before the tape reaches the record head. (See also *Bulk Eraser*)

Extra Play—Also called "long play" or "extended play." Refers to tape that gives more than standard playing time on a standard reel because it employs a thinner base together with a thinner but usually more responsive oxide coating, and thus more tape can fit on a reel.

Fast Forward—The provision on a tape recorder permitting tape to be run rapidly through it in the normal play direction, usually for search or selection purposes.

Feed Reel—Also called "stock," "supply," or "storage" reel. The reel on a tape recorder from which the tape is taken as the machine records or plays.

Flutter—Very short, rapid variations in tape speed, causing pitch and volume variations that were not present in the original sound. A form of distortion.

Four-Channel Sound—Sound produced by four loudspeakers, each being fed a different signal. At present four-channel tape machines are equipped with special heads and electronics that enable them to play back and record four tracks at a time.

Four-Track or Quarter-Track Recording—The arrangement by which four different channels of sound may be recorded on quarter-inch-wide audio tape. These may be recorded as four separate and distinct tracks (monophonic) or two related (stereo) pairs of tracks. By convention, tracks 1 and 3 are recorded in the "forward" direction of a given reel, and tracks 2 and 4 are recorded in the "reverse" direction. (See also *Four-Channel Sound.*)

Frequency—The repetition rate of cyclic energy, such as sound or alternating electrical current, expressed in cycles per second (hertz or Hz) or thousands of cycles per second (kilohertz or kHz). By convention, "bass" frequencies in music extend from about 20 to about 200 Hz. "Treble" sounds are at the high-frequency extreme of the sound spectrum and may extend from 2 or 3 kHz to the frequency limit of audibility (about 18 to 20 kHz). "Middle" (or mid-range) frequencies occupy the remainder of the spectrum, from 200 Hz to about 3 kHz.

Frequency Range—The span between the highest and lowest pitched sounds that a tape recorder or other sound-system component can reproduce at a usable output or volume level.

Frequency Response—Always specified as a range, such as 50 to 15,000 Hz; but in order to be meaningful it must be further defined in terms of decibel variation from absolute flatness over a specified frequency range (e.g., ±3 db from 50 to 15,000 Hz). An indication of a sound system's ability to reproduce all audible frequencies supplied to it, maintaining the original balance among the low, middle (or mid-range), and high frequencies.

Full-Track Recording—Applies to quarter-inch-wide (or less) tape only. It defines track width as essentially equal to tape width.

Gain—The voltage ratio of the output level to the input level for a system or component of a system. Usually expressed in decibels.

Gap—The effective distance between opposite poles of a magnetic head, measured in microinches or microns. Especially critical for playback heads in which gaps must be narrow in order to resolve (reproduce) high-frequency (short wave-length) signals. Recording heads generally have wider gaps than reproducing heads.

Harmonic Distortion—Distortion characterized by the appearance in the output signal of spurious harmonics of the fundamental frequency. Usually expressed as a percentage of the output signal.

Harmonics—Overtones that are integral multiples of the fundamental frequency. In properly balanced a.c.-biased tape recorders, only the odd-order harmonics (primarily the third) are generated by the recording process and these are very low in amplitude.

Head—In a magnetic-tape recorder, the generally ring-shaped electromagnet across which the tape is drawn. Depending on its function, it either erases a previous recording, converts an electrical signal to a corresponding magnetic pattern and impresses it on the tape (record function), or picks up a magnetic pattern already on the tape and converts it to an electrical signal (playback function). Most home recorders have a separate erase head, but combine the record and play functions in a single unit. Professional machines and those intended for the serious amateur have separate heads for erase, record, and playback.

Head Demagnetizer or Degausser—A device used to neutralize possible residual or induced magnetism in heads or tape guides.

Unless the recorder has an automatic head-demagnetizing circuit and nonmagnetic tape guides, periodic use of a head demagnetizer may be necessary to avoid addition of hiss noise to, or even partial erasure of, prerecorded tapes.

Hiss—A high sibilant sound, most often found in tape recording or tape playback. The better the tape system, the lower the hiss.

Hz—Standard abbreviation for Hertz, which is equivalent to "cycles per second" and is the measure of sound frequency or pitch.

Index Counter—An odometer type of counter that indicates revolutions (not feet of tape), usually of the supply reel, thereby making it possible to index selections within a reel of tape and readily locate them later on a given machine.

Input—The terminals, jack, or receptacle provided for the introduction of the electrical input signal voltage into an amplifier or other electronic component.

Input Signal—An electrical voltage embodying the audio information that is presented to the input of an amplifier, tape recorder, or other electronic component.

Intermodulation Distortion—Distortion that results when two or more pure tones produce new tones with frequencies representing the sums and differences of the original tones and their harmonics.

ips—Abbreviation for tape speed (inches per second).

Jack—Receptacle for a plug connector leading to the input or output circuit of a tape recorder or other piece of equipment. A jack matches a specific plug.

kHz—Abbreviation for kilohertz, or one thousand cycles per second. For example, 19 kHz equals 19,000 Hz.

Leader and Timing Tape—Special tough nonmagnetic tape that can be spliced to either end of a magnetic tape to prevent its damage and possible loss of recorded material. Either white or in colors, it usually has some type of marking that enables it to be used as a timing tape. It therefore can be spliced between musical selections to provide desired pauses in playback.

Level Indicator—A device on a tape recorder for indicating the level at which the recording is being made; it serves as a warning against under- or over-recording. It may be a light indicator or a meter. (See also *VU Meter*.)

Low-Noise Tape—Magnetic tape with a signal-to-noise ratio 3 to 5 db better than conventional tapes, making it possible to record sound (especially wide-frequency-range music) at reduced tape speeds without incurring objectionable background noise (hiss) and with little compromise of fidelity. Additional characteristics of most low-noise tapes include extremely good high-frequency sensitivity and a heavy-duty binder system for reduced ruboff of magnetic oxide and an increase in wear life over ordinary tapes.

Low-Print Tape—Special magnetic recording tape significantly less susceptible to print-through (the transfer of signal from one layer of tape to another), which results when tape is stored for long periods of time. These tapes are especially useful for "master recording" (making an original recording from which copies will be made) on professional-quality equipment.

Mil—One one-thousandth of an inch. Tape thickness is usually measured in mils.

Mixer—A device that allows two or more signal sources to be blended, balanced, and fed simultaneously into a tape recorder or amplifier.

Monitor Head—A separate playback head on some tape recorders that makes it possible to listen to the material on the tape an instant after it is recorded and while the recording is still in progress.

Monophonic—Refers to single channel sound as distinguished from stereophonic (two-channel) or quadraphonic (four-channel) sound.

NAB Curves—Standard tape-recorder playback equalization curves established by the National Association of Broadcasters. (See also *Equalization*.)

Noise—Unwanted electrical signals produced by electronic equipment, and rough or nonhomogeneous oxide coatings on magnetic tape.

Open Reel—Tape systems that use reels of tape. To start the tape, it must be threaded by hand from the full to the empty (or takeup) reel.

Oxide—The ferro-magnetic particles which, when properly dispersed in a plastic binder and coated on a backing or base, form the magnetic portion of magnetic tape. Conventional oxide particles are chemically known as gamma ferric oxide, are brown in color, acicular (needlelike) in shape, and of micron length. Less conventional oxides have been developed that exhibit significantly different magnetic properties (and size).

Pause Control—A feature of some tape recorders that makes it possible to stop the movement of tape temporarily without switching the machine from "play" or "record."

Playback—The reproduction of sound previously recorded on a tape. The opposite of *record.*

Playback Head—Magnetic head used to pick up a signal from a tape. Often the same head as is used for recording, but with its circuits changed by means of a record/play switch which also energizes the erase head. (See also *Head.*)

Polyester Base—A plastic-film backing for magnetic tape used for special purposes where strength and resistance to temperature and humidity change are important. (Mylar is a du Pont trade name for their brand of polyester.)

Prerecorded Tape—Tape recordings that are commercially available and generally embody the same material that is available on phonograph records.

Pressure Pad—A device that forces tape into intimate contact with the head gap, usually by direct pressure at the head assembly. Felt or similar material, occasionally protected with self-lubricating plastic, is used to apply pressure uniformly and with a minimum of drag on the backing (non-coated) side of the tape.

Pressure Roller—Also called "pinch roller" or "capstan idler." A hard-rubber roller that holds the magnetic tape tightly against the capstan, permitting the latter to draw the tape off the supply reel and past the hands at a constant speed. (See also *Capstan.*)

Print-Through—Undesired transfer of magnetic pattern from layer to layer of tape on a reel. In most cases, will make recording unusable.

Quadraphonic—Quad for short—refers to four-channel sound.

Reel-to-Reel—Designates those tape machines that do not use a cartridge or cassette. (See also *Open Reel.*)

Rewind Control—A button or lever for rapidly rewinding tape from the takeup reel to the supply reel.

Saturation—The condition reached in magnetic tape recording where output does not increase with increased input, and hence distortion increases significantly.

Separation—The degree to which two stereo signals are kept apart. Stereo realism depends on the successful prevention of their mixture in all parts of a hi-fi or tape system. Tape systems have separation capability superior to that of disk systems.

Signal-to-Noise Ratio—The voltage ratio, usually expressed in decibels, between the loudest undistorted tone recorded and reproduced by the recorder and the noise reproduced when the audio signal is reduced to zero.

Sound-on-Sound—A method by which material previously recorded on one track of a tape may be rerecorded on another track while simultaneously adding new material to it.

Splicing Block—A metal or plastic device incorporating a groove within which ends of the tape to be spliced are held. An additional diagonal groove provides a path for a razor blade to follow in cutting the tape. It makes splices very accurately using narrow-width (7/32″) splicing tape. (See also *Tape Splicer.*)

Splicing Tape—A special pressure-sensitive, non-magnetic tape used for joining two lengths of magnetic tape. Its "hard" adhesive will not ooze, and consequently will not gum up the heads or cause adjacent layers of tape on the reel to stick together.

Squeal—The audible noise caused by alternate sticking and release of tape. It may occur at heads, pressure pads, or guides where friction develops with the face or back side of a magnetic tape. It is largely eliminated by regular cleaning of suspected surfaces.

Takeup Reel—The reel on the tape recorder that accumulates the tape as it is recorded or played.

Tape Guides—Grooved pins or rollers mounted between and at both sides of the tapehead assembly to position the magnetic tape correctly on the head as it is being recorded or played.

Tape Lifters—A system of movable guides that automatically prevents the tape from contacting the recorder's heads during fast-forward or rewind modes of operation, thus preventing head wear.

Tape Monitoring—See *Monitor Head*

Tape Player—A unit that is not capable of recording and is used only for playing prerecorded tapes.

Tape Speed—The speed at which tape moves past the head in recording or playback modes. Standard tape speed for home use is 7½ ips or half that speed (3¾ ips). Speeds of 1⅞ and $1\frac{5}{16}$ ips are found on some machines, but on reel-to-reel recorders are usually suitable only for non-critical voice recording. Some cassette and cartridge machines, using special tape and circuits, achieve very good results at the slow speeds. Professional recording speed (for making original master tapes of music, for example) is usually 15 ips and sometimes higher.

Tape Splicer—A device, similar to a film splicer, for splicing magnetic tape automatically or semi-automatically. Different models vary in operation, most using splicing tape; some professional units employ heat. (See also *Splicing Block.*)

Tape-Transport Mechanism—The platform or deck of a tape recorder on which the motor (or motors), reels, heads, and controls are mounted. It includes those parts of the recorder other than the amplifier, preamplifier, loudspeaker, and case.

Telephone Pickup—Any of several devices used to feed telephone conversations into a tape recorder, usually without direct connection to the telephone line and operating by magnetic coupling.

Tensilized Polyester—A polyester tape backing that has been prestretched principally in the lengthwise direction to increase resistance to further stretching.

Tone Controls—Control knobs on a tape-recorder amplifier used to vary bass and treble response to achieve the most desirable balance of tone during playback.

Track—The path on the magnetic tape along which a single channel of sound is recorded.

Two-Track Recording—On quarter-inch-wide tape, the arrangement by which only two channels of sound may be recorded, either as a stereo pair in one direction or as separate monophonic tracks (usually in opposite directions).

Uniformity—In terms of magnetic tape properties, a figure of merit relating to the tape's ability to deliver a steady and consistent output level when being recorded with a constant input. Usually expressed in decibel variation from average at a mid-range frequency.

VU Meter—A "volume unit" meter that indicates audio-frequency levels in decibels relative to a fixed 0-db reference level. The meter movement differs from those of ordinary voltmeters in that it has a specified response adapted to monitoring speech and music. Used in many home and most professional recorders to monitor recording levels and maintain them within the distortion limits of the tape.

Wave Length—In tape recording (and referring specifically to the tape magnetization created by pure single-tone recording), the shortest physical distance between two peaks of the same magnetic polarity.

Wow—A form of distortion in sound-reproducing systems caused by relatively slow periodic variation in the speed of the medium (such as tape) and characterized by its effect on pitch.

INDEX

Figures in *italic* type refer to illustrations.

About the Author

HANS FANTEL, a former editor of *Stereo Review,* is one of the best-known writers on audio. His book *The True Sound of Music: A Practical Guide to Sound Equipment for the Home* (also published by E. P. Dutton & Co.) has been singled out by the *Library Journal* as one of the best publications in its field and has been a selection of the Popular Science and Playboy Book clubs. He also contributes articles on audio to *The New York Times, Popular Science, Popular Mechanics,* and *Opera News* and serves as consultant on audio to the Reader's Digest Record Club.

Mr. Fantel also writes on historical subjects. His book *The Waltz Kings* is a history of the Viennese waltz and its chief creator, Johann Strauss. His most recent biographical work is *William Penn–Apostle of Dissent,* dealing with the role of the Quakers in early American history.

Mr. Fantel, whose books have appeared in nine languages, makes his home in New York City and the Berkshire Hills of Massachusetts.